Springer Transactions in Civil and Environmental Engineering

Springer Transactions in Civil and Environmental Engineering (STICEE) publishes the latest developments in Civil and Environmental Engineering. The intent is to cover all the main branches of Civil and Environmental Engineering, both theoretical and applied, including, but not limited to: Structural Mechanics, Steel Structures, Concrete Structures, Reinforced Cement Concrete, Civil Engineering Materials, Soil Mechanics, Ground Improvement, Geotechnical Engineering, Foundation Engineering, Earthquake Engineering, Structural Health and Monitoring, Water Resources Engineering, Engineering Hydrology, Solid Waste Engineering, Environmental Engineering, Wastewater Management, Transportation Engineering, Sustainable Civil Infrastructure, Fluid Mechanics, Pavement Engineering, Soil Dynamics, Rock Mechanics, Timber Engineering, Hazardous Waste Disposal Instrumentation and Monitoring, Construction Management, Civil Engineering Construction, Surveying and GIS Strength of Materials (Mechanics of Materials), Environmental Geotechnics, Concrete Engineering, Timber Structures.

Within the scopes of the series are monographs, professional books, graduate and undergraduate textbooks, edited volumes and handbooks devoted to the above subject areas.

The proposal for each volume is reviewed by the main editor and/or the advisory board. The chapters in each volume are individually reviewed single blind by expert reviewers (at least four reviews per chapter) and the main editor.

Mohit Somani • Manoj Datta • G. V. Ramana
Editors

Landfill Mining of Legacy MSW: Global Insights and Indian Experience

Springer

Editors
Mohit Somani
School of Infrastructure
Indian Institute of Technology Bhubaneswar
Khurda, Odisha, India

Manoj Datta
Department of Civil Engineering
Indian Institute of Technology Delhi
New Delhi, Delhi, India

G. V. Ramana
Department of Civil Engineering
Indian Institute of Technology Delhi
New Delhi, Delhi, India

ISSN 2363-7633 ISSN 2363-7641 (electronic)
Springer Transactions in Civil and Environmental Engineering
ISBN 978-981-95-3958-1 ISBN 978-981-95-3959-8 (eBook)
https://doi.org/10.1007/978-981-95-3959-8

This Springer imprint is published by the registered company Springer Nature Singapore Pte Ltd.
The registered company address is: 152 Beach Road, #21-01/04 Gateway East, Singapore 189721, Singapore

Foreword

The challenge of managing legacy municipal solid waste (MSW) is one of the most urgent yet underexplored issues in the pursuit of sustainable urban development. As cities worldwide grapple with overflowing dumpsites, deteriorating environmental quality, and pressing circular economy mandates, landfill mining has emerged as both a necessity and an opportunity. It is in this context that the book *Landfill Mining of Old MSW Dumps: Global Insights and Indian Experience* makes a timely and invaluable contribution.

This book distinguishes itself in several ways. It is among the rare works that seamlessly bridge global perspectives with the rapidly evolving Indian experience. The initial chapters provide an overview of international practices, including Europe's early initiatives and specific lessons from Italy, before carefully situating them against the distinctive trajectory that landfill mining has taken in India since 2019. Such a comparative lens enables readers to appreciate both the common challenges of economic feasibility and the unique adaptations driven by scale, urgency, and regulatory frameworks in India.

This book moves beyond high-level discussions to offer a detailed analysis of the material fractions recovered through landfill mining. With contributions from leading experts across institutions, the chapters present rigorous characterization of gravel and soil-like material, combustibles, and residual fractions. Importantly, this book does not shy away from confronting difficult truths, especially the risks associated with the widespread assumption of inertness in fine fractions. By addressing contaminants of concern, microplastics, and long-term geotechnical and environmental implications, it advocates for precautionary and scientifically informed reuse pathways.

The inclusion of a checklist-based framework for project planning and evaluation reflects the editors' commitment to producing a work that is not only academically sound but also practically applicable. Case studies and field investigations further enhance its value as a reference for practitioners, engineers, regulators, and students alike.

In essence, this book equips readers with the knowledge to critically evaluate landfill mining as both a remediation tool and a resource recovery strategy. It underscores that while landfill mining holds immense promise, its success depends on risk-aware, technically robust, and context-sensitive practices. For anyone concerned with waste management, environmental protection, or sustainable infrastructure, this book is not just worth reading but knowledge learning that would change the landscape of urban development and revitalization of land use from many urban dumpsites.

James Law
International Landfill Expert and Consultant
Virginia, United States

Contents

Contributors

M. Ambal Devi Department of Civil Engineering, National Institute of Technology Tiruchirappalli, Tiruchirappalli, Tamil Nadu, India

G.L. Sivakumar Babu Department of Civil Engineering, Indian Institute of Science Bangalore, Bengaluru, Karnataka, India

Munish Kumar Chandel Environmental Science and Engineering Department, Indian Institute of Technology Bombay, Powai, Mumbai, Maharashtra, India

Manoj Datta Department of Civil Engineering, Indian Institute of Technology Delhi, New Delhi, India

Walter Giacetti Ecoambiente srl, Sarzano, RO, Italy

Deepak Kumar Haritwal Department of Civil Engineering, Indian Institute of Technology Delhi, New Delhi, India

Vaishnavi Jahagirdar Department of Civil Engineering, Indian Institute of Technology Guwahati, Guwahati, Assam, India

Ajay S. Kalamdhad Department of Civil Engineering, Indian Institute of Technology Guwahati, Guwahati, Assam, India

Arif Khan Department of Civil Engineering, Indian Institute of Technology Guwahati, Guwahati, Assam, India

James Law International Landfill Expert and Consultant, Virginia, United States

Roshan Vilasrao Mankhair Environmental Science and Engineering Department, Indian Institute of Technology Bombay, Powai, Mumbai, Maharashtra, India

Anil Kumar Mishra Department of Civil Engineering, Indian Institute of Technology Guwahati, Guwahati, Assam, India

Supriya Mohanty Department of Civil Engineering, Indian Institute of Technology (BHU), Varanasi, India

Kasinathan Muthukkumaran Department of Civil Engineering, National Institute of Technology Tiruchirappalli, Tiruchirappalli, India

Anusree Nalladiyil Centre for Sustainable Technologies, Indian Institute of Science Bangalore, Bengaluru, Karnataka, India

Debaprakash Parida Department of Civil Engineering, Indian Institute of Technology Delhi, New Delhi, India

B. Prathima Centre for Sustainable Technologies, Indian Institute of Science Bangalore, Bengaluru, India

Department of Civil Engineering, B M S College of Engineering, Bengaluru, India

Roberto Raga ICEA, Department of Civil, Environmental and Architectural Engineering, University of Padua, Padova, Italy

G.V. Ramana Department of Civil Engineering, Indian Institute of Technology Delhi, New Delhi, India

Parul Rawat Department of Civil Engineering, Indian Institute of Technology (BHU), Varanasi, India

René Møller Rosendal Danish Waste Solutions ApS, Copenhagen, Denmark

AV Miljø Landfill, Copenhagen, Denmark

Ankur Saini MVW Lechtenberg & Partner, Duisburg, Germany

Manish Sharma School of Infrastructure, Indian Institute of Technology Bhubaneswar, Khurda, Odisha, India

Mohit Somani School of Infrastructure, Indian Institute of Technology Bhubaneswar, Khurda, Odisha, India

Paula Tan Department of Civil, Environmental and Architectural Engineering, University of Padova, Padova, Italy

Abhay Kumar Verma Indian Institute of Technology—BHU, Varanasi, Uttar Pradesh, India

Enrico Villani Ecoambiente srl, Sarzano, RO, Italy

Part I
Global and Indian Scenario

Chapter 1
Global Scenario of Landfill Mining

René Møller Rosendal

1.1 Introduction

Landfill Mining (LFM), Landfill Mining and Reclamation (LFMR), Enhanced Landfill Mining (ELFM), and Biomining are terms frequently used to describe broadly similar practices with minor variations in scope or emphasis. Fundamentally, these approaches share the same objective: excavating and processing waste previously disposed of in dumpsites or sanitary landfills in order to recover resources, reclaim land, and reduce the environmental impacts and health risks associated with legacy waste.

Although the concept dates back decades, it continues to evolve as a waste management strategy. More than 120 projects have been recorded since the first documented case in 1953 (Rosendal 2009), but comprehensive evaluations remain scarce. Well-documented case studies with detailed material analyses, cost-benefit assessments, and standardized evaluation methods are still rare (Krook et al. 2012). Consequently, while LFM has been a subject of research for over 20 years, the field remains in an early stage of maturity without universally accepted methodologies or performance benchmarks.

Global strategies for LFM share common principles, yet their outcomes differ greatly depending on regional variations in waste composition, recycling opportunities, market demand, and regulatory frameworks. This chapter aims to outline the conceptual foundation of landfill mining, discuss the associated challenges, and summarize lessons learned from EU-based projects that have demonstrated notable success.

R. M. Rosendal (✉)
Danish Waste Solutions ApS, Copenhagen, Denmark

AV Miljø Landfill, Copenhagen, Denmark
e-mail: rmr@danws.dk

M. Somani et al. (eds.), *Landfill Mining of Legacy MSW: Global Insights and Indian Experience*, Springer Transactions in Civil and Environmental Engineering, https://doi.org/10.1007/978-981-95-3959-8_1

1.2 Why Landfill Mining?

The disposal of human waste materials in designated areas is as old as civilization itself. Archaeological evidence reveals that even prehistoric communities accumulated refuse such as shells, bones, and broken tools in designated sites. Later, during the Iron Age, by-products like metal slag from blacksmiths added to such deposits. These remnants now serve as valuable records of past societies.

In modern times, landfilling remains one of the most widespread waste management practices worldwide due to its simplicity and relatively low cost compared with alternatives such as incineration, anaerobic digestion, or composting. However, burying waste has profound environmental consequences. Even engineered sanitary landfills release greenhouse gases, particularly methane (CH_4) and carbon dioxide (CO_2), both of which accelerate climate change. The International Solid Waste Association (ISWA 2016) estimated that by 2025, landfills could contribute around 10% of global greenhouse gas emissions.

The scale of the issue is immense: global municipal solid waste generation amounts to approximately 2.01 billion tons annually, with roughly one-third of it managed unsafely. Around 40% of this waste ends up in open dumpsites that directly serve billions of people, especially in regions with limited formal waste collection systems. For instance, open dumping accounts for up to 90% of disposal in many African countries and 45% in Latin America and the Caribbean (UNEP 2021).

Against this backdrop, LFM emerges as both an innovative and sustainable approach. It not only mitigates environmental risks by reducing methane emissions and leachate generation but also enables the recovery of valuable resources. Furthermore, it contributes to the broader objectives of a circular economy, where waste is viewed as a resource rather than a liability. By reclaiming land and minimizing ecological footprints, LFM provides a compelling option for addressing the legacy of uncontrolled waste disposal.

1.3 Pros and Cons—Mining Landfills

Historically, LFM initiatives were primarily motivated by the need to address immediate problems, such as limited landfill capacity or local pollution concerns. More recently, however, projects have begun integrating resource recovery objectives, including the extraction of soil-like fine fractions for cover material and the production of refuse-derived fuel (RDF) for waste-to-energy plants. Although recycling has often been considered secondary, this trend is gradually shifting toward a stronger focus on material recovery.

The technologies employed in landfill mining vary from conventional earthmoving equipment—such as excavators, bulldozers, conveyors, and screening units—to more advanced tools. Traditional methods have shown only moderate success in producing marketable secondary materials. To improve recovery efficiency, new technologies are being introduced, including advanced screening systems, sensor-based sorting (e.g., infrared and X-ray technologies), and geospatial monitoring tools for site characterization and real-time assessment.

Growing awareness of resource scarcity and sustainability, combined with recent advances in technology, has prompted a re-evaluation of landfills. They are increasingly recognized as "urban mines," with the potential to supply raw materials that reduce dependence on virgin resources. This shift in perspective is reshaping global waste management strategies and driving research on both environmental and economic dimensions of LFM.

Several global factors have contributed to the rapid growth and persistence of landfills as a dominant waste management practice:

- Urban population growth: Expanding cities generate vast amounts of waste, creating constant demand for disposal facilities.
- Insufficient recycling and recovery systems: In many regions, limited infrastructure and weak enforcement mean that recyclable and reusable materials continue to end up in landfills.
- Economic considerations: Compared to alternative waste management methods, landfilling often remains the least expensive option, encouraging its continued use.
- Limited awareness of long-term impacts: Historically, the environmental and health consequences of landfills such as groundwater contamination and greenhouse gas emissions have been underestimated or overlooked.
- Land availability: In less densely populated areas, the relative abundance of land has facilitated the expansion of landfill sites, further entrenching their use.

While environmental benefits are often emphasized, landfill mining also carries notable economic advantages:

- Reduced disposal costs: By recovering and processing previously landfilled materials, the overall waste volume requiring management decreases, leading to cost savings for municipalities and waste authorities.
- Revenue from secondary resources: Extracted materials such as metals, plastics, and aggregates can be reintroduced into recycling markets, creating potential revenue streams.
- Redevelopment opportunities: Once reclaimed, former landfill sites may be repurposed for infrastructure, commercial, or community projects, adding further economic value to the process.

Taken together, these factors make landfill mining an attractive proposition for countries grappling with resource scarcity, rising waste generation, and mounting environmental pressures. It offers a pathway toward resource efficiency, environmental restoration, and economic gain. However, successful implementation requires careful attention to several obstacles. Both technological limitations such as the efficiency of material recovery systems and economic barriers, including fluctuating market demand for secondary resources, must be addressed. Without tackling these issues, landfill mining projects risk falling short of their potential and failing to gain public or institutional acceptance.

1.4 The Concept of Landfill Mining

Implementing LFM projects requires careful planning and rigorous management at every stage. Before excavation can begin, a series of preliminary investigations must be carried out to secure the necessary environmental permits and ensure compliance with regulatory frameworks. A comprehensive overview of the entire process is essential, as it allows practitioners to anticipate operational challenges and integrate environmental, technical, and social considerations into the project design (Rosendal 2009; Rosendal et al. 2017).

Table 1.1 summarizes the main operational stages involved in LFM, outlining the requirements of each stage and the corresponding support systems needed to manage them effectively (Rosendal 2024).

Table 1.1 Planning for LFM excavation and disposal (Rosendal 2024)

Unit operation	Operational requirement	Support systems
Pre-operational period	Environmental impact assessment Baseline monitoring Desktop studies/interviews Stakeholder and community consultation Environmental permitting	Documentation and monitoring systems Engagement mechanisms
Site preparation	Establish site services, access routes, and working platforms Temporary storage and disposal facilities Security measures	Environmental safeguards Monitoring programs Health and safety plan
Excavation	Define excavation rate and material flow Identify waste types to be managed Plan for segregation, separation, and dewatering Transport logistics and required machinery	Record-keeping procedure
Materials handling	Manage variations in waste volume and composition Segregation and dewatering processes Transport systems and equipment requirements	Environmental monitoring Record management
Replacement	Define placement method Equipment requirements Safeguards to prevent pollutant migration	Environmental protection and monitoring Record-keeping
Final disposal off-site	Transport arrangements and logistics	Record management
Final disposal on-site	Characterize designated disposal areas Quantify materials for replacement Engineering works for site preparation Placement operations and restoration timelines	Environmental protection Monitoring and safety oversight Record-keeping
Post-treatment management	Validation of excavated areas Long-term monitoring of on-site disposal areas Maintenance reviews	Data collection and record management

1.5 Material and Waste Composition

One of the most frequently investigated aspects of LFM is the characterization of deposited waste. Understanding the composition of landfilled material is essential for assessing recovery potential, designing appropriate excavation strategies, and determining environmental risks.

In general, excavations of municipal landfills reveal the following approximate distribution of materials:

- 50–60% soil-like fractions, consisting of cover material and highly decomposed waste;
- 20–30% combustible fractions, such as plastics, paper, textiles, and wood;
- ~10% inert or inorganic fractions, including glass, concrete, stones, and ash;
- 1–2% metals, predominantly ferrous scrap (Krook et al. 2018).

The proportion of hazardous materials is typically small, often reported at below 1–2%, though this varies considerably across sites. The recovery potential of a landfill is influenced not only by its waste composition but also by its physical and chemical conditions, as well as the efficiency of excavation and processing technologies (Cossu et al. 1996).

A major factor affecting waste characterization is the soil-to-waste ratio, which can differ widely among landfills. This variation is linked to several factors: the extent of daily and final cover layers, the degree of waste compaction, the age of the deposited material, local climatic and moisture conditions, and the screen size applied during sorting operations. Reported soil-to-waste ratios range anywhere between 20:80 and 75:25 (Tammemagi 1999).

While significant volumes of soil-like material can be recovered during LFM operations, its potential reuse is contingent on chemical safety. Investigations have revealed contamination from hydrocarbons such as mineral oils, as well as polycyclic aromatic hydrocarbons (PAHs), volatile organic compounds (VOCs), heavy metals, pesticides, and polychlorinated biphenyls (PCBs) (Geusebroek 2001; Hull et al. 2001). Such findings highlight the importance of rigorous testing and compliance with national or regional environmental standards before reclaimed soil can be applied for beneficial use.

At present, Europe lacks unified criteria for assessing the quality of excavated soils and waste fractions. Instead, requirements for reuse remain governed by individual national regulations, creating discrepancies across the EU and complicating the development of harmonized approaches (Rosendal 2009).

1.6 Economics of Landfill Mining

LFM has the potential to reduce or even eliminate the costs typically associated with landfill closure, while at the same time alleviating long-term environmental risks. However, the economic viability of LFM is highly context-specific and influenced by a range of technical, financial, and market-related factors.

One of the key determinants is the depth of waste deposits and the soil-to-waste ratio. Deeper landfills require more extensive excavation, which increases operational costs, while high volumes of cover soil relative to waste may reduce the recovery potential of valuable fractions (Rosendal 2015; Hyks et al. 2016).

The economic benefits of LFM are often expected to come from the recovery of recyclable materials. These benefits, however, are conditional on multiple factors, including the quality and purity of the recovered fractions, local demand for secondary materials, and prevailing market prices. Under favorable conditions, valuable fractions such as ferrous and non-ferrous metals (e.g., aluminum), plastics, glass, and fines can generate revenue streams. Such opportunities are particularly evident in specialized contexts, such as industrial landfills or those linked to sectors like car shredding and metal scrap industries. Conversely, landfills containing toxic or complex industrial residues—such as those associated with historical glass or battery production—can be extremely costly to remediate, often outweighing any potential material recovery revenues.

Although there may be hundreds of thousands of candidate sites worldwide suitable for reclamation, LFM remains underutilized. This is primarily due to limited site-specific information and the complexity of conducting robust economic evaluations (Laner et al. 2019; Rosendal et al. 2017).

To properly assess feasibility, a comprehensive economic evaluation is essential. Such assessments must go beyond short-term recovery revenues and account for a wide range of cost-saving and value-adding factors, including:

- The reduction or elimination of landfill capping expenses, long-term monitoring, and post-closure maintenance;
- The potential avoidance of remediation costs linked to pollution control;
- Optimization of machinery use and logistics during excavation;
- Increases in the market value of reclaimed land, especially where redevelopment opportunities exist;
- Cost savings associated with avoiding the development of new landfill infrastructure, particularly if the reclaimed land can be repurposed for waste disposal (Rosendal 2015).

Ultimately, the costs and benefits of LFM vary considerably across projects, depending on site-specific conditions, project objectives, and the local economic context. Table-based or framework evaluations of capital investments, operational expenses, and expected benefits are therefore necessary to determine whether reclamation is financially viable in a given case (Rosendal 2015) (Table 1.2).

The financial performance of LFM projects is shaped by a wide range of site-specific and external factors. A thorough assessment is therefore essential before project initiation. Key aspects to be analyzed when evaluating the economic feasibility include:

- Remaining landfill capacity and projected waste generation;
- Closure or expansion costs if the existing landfill reaches its capacity;

Table 1.2 Capital and operational costs related to LFM (Rosendal 2024)

Capital costs	Operational costs
Preliminary investigations (e.g., site surveys, environmental assessments)	Labor for daily operations (e.g., excavation, sorting, and materials handling)
Site preparation works (access roads, working platforms, storage areas)	Fuel, energy, and routine maintenance of equipment
Acquisition or rental of excavation, reclamation, and sorting machinery	Administrative and regulatory compliance costs (e.g., permits, documentation)
Procurement of worker safety equipment and protective systems	Training programs for workers on health, safety, and equipment operation
Construction or upgrading of materials handling and processing facilities	Transportation and hauling costs for recovered or residual materials
Acquisition or rental of transport and hauling vehicles	

- Future liabilities, such as long-term monitoring, remediation, and potential environmental damages;
- Land value after reclamation, which may allow for redevelopment or alternative uses;
- Market demand for recyclables and recovered secondary resources (Hyks et al. 2016).

One of the most significant benefits of LFM is the extension of the operational lifespan of existing landfill sites. By recovering and processing deposited materials, space is freed up, which delays the need to locate, design, permit, and construct new facilities a process that is costly, time-consuming, and often contested by local communities.

Economic assessments must also incorporate environmental costs and benefits, which are often more difficult to quantify. Conventional financial tools such as Net Present Value (NPV), Benefit-Cost Ratio (BCR), or Internal Rate of Return (IRR) can be applied, but they require the inclusion of externalities. For instance, avoided methane emissions represent a clear climate benefit, yet translating this into monetary terms remains complex. Similarly, improvements in local environmental quality such as reduced smoke from open burning, mitigation of odors and fly nuisance, and decreased groundwater contamination from leachate are rarely monetized in practice, despite their clear social and health benefits.

In general, the benefits of LFM can be grouped into two broad categories:

- Operational benefits, such as extending landfill life and reducing future liabilities.
- Resource recovery benefits, including revenues from recyclables and increased land value after reclamation.

Conversely, costs are typically divided into capital investments (e.g., site preparation, equipment purchase, safety systems) and operational expenses (e.g., labor, maintenance, compliance, and transport). In practice, however, the distinction is not always clear-cut. For example, worker training in safety procedures may be considered a one-time capital cost but can also represent an ongoing operational expense if regular retraining is required to maintain compliance and safety standards.

Given the high financial commitment and inherent risks, accurate economic evaluation is critical prior to launching an LFM initiative. This evaluation must be informed by detailed investigations conducted in advance of excavation, including waste characterization studies, environmental risk assessments, and an understanding of the local regulatory and planning context (Rosendal 2009; Laner et al. 2019).

1.7 Landfill Mining in Europe

Across Europe, there are more than half a million landfills, most of which are older, non-sanitary municipal solid waste facilities that were constructed without modern environmental safeguards (Vollprecht et al. 2021). These landfills vary greatly in terms of age, design, and waste composition. Compared to other parts of the world, European sites present unique challenges for LFM, with further differences even within the continent. For example, projects in Northern Europe often face difficulties in achieving economic viability, primarily due to high labor and contractor costs.

Although the theoretical potential of LFM in Europe appears substantial, practical implementation is hindered by technological, economic, and regulatory constraints (Krook et al. 2012). In response to these barriers, the EURELCO was launched in 2014, bringing together 38 organizations from 13 EU member states, including research institutions, industry, and governmental bodies. Its objective is to create a collaborative platform to address the technological, economic, social, legal, and environmental dimensions of LFM. The concept promoted by the consortium ELFM focuses on recovering secondary raw materials and energy resources from landfills in a safe and environmentally responsible manner.

Despite strong interest, large-scale, real-world examples remain scarce. Among the most notable initiatives are Belgium's Closing-the-Circle project and the EU-funded NEW-MINE training network (Krook 2021). Both have shown encouraging results, particularly in upcycling waste into marketable high-quality materials. However, they also highlight the need for further investment to increase the technology readiness level of current approaches.

Two additional examples of successful EU projects include the Sandford Farm Landfill (UK) and the Samaritaine Landfill (France).

1.7.1 Sandford Farm Landfill, UK

The Sandford Farm site, located near Reading, covered around 20 hectares and was originally a licensed industrial and commercial landfill without lining or environmental controls. Although groundwater contained elevated levels of metals, hydrocarbons, and nitrogen, severe contamination was not observed (Beese-Vasbender et al. 2021). After restoration in 2005, plans for residential redevelopment faced delays due to public opposition but were finally approved in 2012. Given its proximity to London, the land carried significant real estate value.

LFM operations began in 2013 and concluded in 2016, with the objective of reclaiming the land for housing. Excavation was conducted in phases, followed by waste sorting and recovery. Techniques included selective excavation for large items, air-knife separation for plastics, and vibratory screening to classify waste into <40 mm, 40–100 mm, and >100 mm fractions. Smaller fractions were immediately reused, while larger fractions required further treatment. Metals were extracted using magnetic separators, and manual sorting was carried out to remove unsuitable materials.

Recycling strategies prioritized on-site reuse: soil-like fractions and aggregates were reused directly or after crushing; plastics were applied in open-space areas; metals (≈370 tons) were sold to recycling facilities; while unsuitable waste such as textiles and timber were either disposed of or reused off-site. Overall, approximately 95% of the waste material was reused.

The redevelopment resulted in the construction of over 460 residential units with appropriate gas protection measures. Importing replacement fill and sending waste to alternative landfills would have cost ~£51 million. By reusing materials on site, costs were reduced to ~£12 million, with scrap metal sales generating ~£26,000. Although the total profit margin for the developer is undisclosed, the project demonstrated the cost-effectiveness of on-site material reuse (Beese-Vasbender et al. 2021).

1.7.2 Samaritaine Landfill, France

The Samaritaine site is located within a sand dune complex in Normandy, covering ~4320 m^2 and designated as part of the Natura 2000 network. Historically, it served as a municipal landfill between 1965 and the 1980s. Due to its coastal location, it was highly vulnerable to erosion and flooding, leading to waste leakage into the estuary over time. By 2016, severe storms had exposed large volumes of waste, posing environmental and tourism-related risks for the region (Beese-Vasbender et al. 2021).

Remediation began in late 2017 and finished in early 2018. The process included vegetation removal, excavation with mechanical shovels, and on-site sorting. Waste was separated into fractions (>100 mm, 6–100 mm, and <6 mm), with metals recovered magnetically and hazardous materials like asbestos isolated. Weather conditions strongly influenced screening performance, with daily throughput ranging from 100 to 550 m^3.

Clean sand recovered from the excavation was reused on-site to reshape dunes and stabilize the area against future erosion. In total, the rehabilitation project cost approximately €1.5 million, and the restored site was repurposed as a natural dune habitat within the protected Natura 2000 network.

1.8 Conclusions

Research on the technical feasibility of resource recovery from landfills has shown that it is indeed possible to upcycle waste and generate high-quality secondary materials. Nevertheless, significant additional investment is required to advance the technological readiness of these processes and to clarify which high-value, marketable products can be reliably extracted. At present, the European market does not provide sufficient economic incentives for large-scale landfill mining, though this situation may shift once the demand for raw materials intensifies and the willingness to invest in secondary resources grows.

From an economic perspective, environmental remediation can be achieved, but typically at considerable expense. To date, only large-scale projects in which reclaimed land has been repurposed for new developments have demonstrated financial viability. This indicates that resource recovery alone is rarely sufficient to justify landfill mining initiatives.

Another limiting factor is the decline of collaborative structures. The EURELCO, once a key platform for advancing research and policy in this field, is unfortunately no longer active. As political and financial support has weakened, momentum has slowed. A renewed interest in landfill mining within Europe is likely to emerge only when critical shortages of secondary materials arise and society is prepared to pay for their recovery.

The discussion on the circular economy is growing in the European Union and their member states but issues with landfilling (eternal aftercare and pollution issues) and landfill mining as a tool to solve the issues is unfortunately not of their concern and on the political agenda.

References

Beese-Vasbender, P., Caterina, D., Ceulemans, M., Chambers, J., Ben Dashwood, B., De Rijdt, R., De Smet, S., Ducheyne, A., Inauen, C., Isunza Manrique, I., Laura Lamair, L., Loisel, S., Moreaux, S., Neculau, C., Nguyen, F., Marta P., Van Raemdonck, M., Vivien, G., Watlet, A., & Wille, E (2021). Landfill miner guide, RAWFILL Interreg North-West Europe EU. Funded by the European Regional Development Fund and co-funded by the Walloon Region

Cossu R, Hogland W, Salerni E (1996) Landfill Mining in Europe and USA. In: ISWA year book, International Solid Waste Association, pp 107–114

Geusebroek HJK (2001) Complete removal of two landfills in Veenendaal (NL). In: Proceedings of Sardinia 91, 8th International Landfill Symposium, vol IV. Cisa Publisher, Cagliari, pp 561–567

Hyks J, Rosendal RM, Ludwigsen K, Hjelmar O (2016) Landfill mining of shredder residues: lessons learnt from Danish case study, special issue: characterizing anthropogenic stocks: methods and applications in resources. Conserv Recycl J

Hull RM, Krogmann U, Storm PF 2001. Characterization of municipal solid waste reclaimed from a landfill. Proceeding of the 8th International Waste Management and Landfill Symposium, 567–576.

ISWA (2016) A roadmap for closing waste dumpsites

Krook J, Svensson N, Eklund M (2012) Landfill mining: A critical review of two decades of research. Waste Manag 32(3):513–520

Krook J (2021a) Enhanced landfill mining, concept and challenges, The Veolia Institute review facts reports—industry and waste: toward the circular economy, 20–21

Krook, J., Cleall, P., Rosendal, R.M., Carvalho, T., (2018). Recovery technologies for materials in landfills deliverable 2.1, COST action mining the European Anthroposphere (MINEA) working group 2: resource potential of waste from landfills

Laner D, Esguerra JL, Krook J, Horttanainen M, Kriipsalu M, Rosendal RM, Stanisavljevic N (2019) Systematic assessment of critical factors for the economic performance of landfill mining in Europe: what drives the economy of landfill mining? Waste Manag 95(2019):674–686

Rosendal RM (2009) Landfill mining, process, feasibility, economy, benefits and limitations. RenoSam

Rosendal RM (2015) The economics of landfill mining—focus on the length of the aftercare. In: Proceedings Sardinia 2015, 15th international waste management and landfill symposium S. Margherita di Pula, Cagliari. 5–9. October, 2015 by CISA Publisher, Italy

Rosendal RM, Christensen JM, Reimann P, Overgaard LB, Poulsen S, Felding P, Sundberg P, Nielsen S (2017) Landfill mining demonstration project at Skaarup Landfill, Denmark, Proceedings Sardinia 2017, 16th International Waste Management and Landfill Symposium S. Margherita di Pula, Cagliari, Italy; 2–6 October, 2017 by CISA Publisher, Italy

Rosendal RM (2024) Chapter 15—landfill mining—opportunities and challenges in a European perspective, solid waste management for resource-efficient systems circularity in action, waste and the environment: underlying burdens and management strategies. In: Singh R, Mohapatra S, Jong M-C (eds) , pp 317–325

Tammemagi H (1999) Landfill mining—Lancaster County's experiment could succeed in Canada. Hazardous materials management. Southam Inc

UNEP (2021) Roadmap for the progressive closure of dumpsites in Latin America and the Caribbean -the Coalition for the closure of dumpsite

Vollprecht D, Machiels L, Jones PT (2021) The EU training network for resource recovery through enhanced landfill mining—a review. Processes 9(2):394

Chapter 2
Case Study of a Successful Landfill Mining Project in Italy

Roberto Raga, Walter Giacetti, Paula Tan, and Enrico Villani

2.1 Introduction

Taglietto landfills are located in northern Italy, approximately 30 km south west of the Venice lagoon, near the small town of Villadose, province of Rovigo. The landfill is composed of four sites (T0, T1, T2, and T3 in Fig. 2.1), in operation between 1980 and 1996, for the disposal of municipal solid waste (MSW) and non-hazardous special waste.

The construction of T0 was approved at the end of the 70s and the disposal activities started at the beginning of 1980. By the time T0 was close to full capacity, toward the end of 1982, T1 was ready to start landfill operations, which went on until the beginning of 1986. In the meantime, the first legislation on landfill design construction had been issued in 1982 in Italy, then completed with a ministerial degree in 1984 providing detailed requirements to be fulfilled for the design and construction of new landfills. For this reason, the additional landfill T2 and T3, built in 1986 and 1992, respectively, have to be considered controlled landfills and they feature, unlike T0 and T1, constructed barriers made of natural and synthetic materials, leachate drainage and biogas collection systems.

The stratigraphy of the area where the landfill is located is outlined (depths in meters below ground level, bgl):

0 m to −2 m: Sandy loams (widely used for agricultural activities)

R. Raga (✉) · P. Tan
ICEA, Department of Civil, Environmental and Architectural Engineering, University of Padua, Padova, Italy
e-mail: roberto.raga@unipd.it

W. Giacetti · E. Villani
Ecoambiente srl, Sarzano, RO, Italy

M. Somani et al. (eds.), *Landfill Mining of Legacy MSW: Global Insights and Indian Experience*, Springer Transactions in Civil and Environmental Engineering, https://doi.org/10.1007/978-981-95-3959-8_2

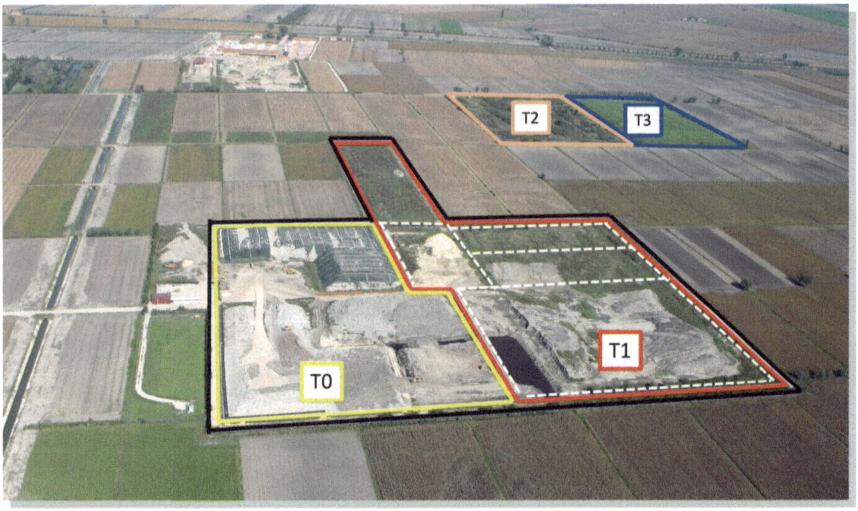

Fig. 2.1 Aerial view of Taglietto landfill during the LFM project, showing the perimeters of T0, T1, T2, and T3 landfill areas. The position of the bentonite clay wall is outlined in black. The white dashed lines outline the position of steel sheet piles installed for mechanical stability during excavation

−2 m to −6 m: Clay and silty clay
−6 m to −8 m: Sand and silty sand (shallow aquifer)
−8 m to −10.5 m: Silty clay layer
−10.5 m to −20 m: Sand and silty sand (deep aquifer)

The clay layer between −8 m and −10.5 m was previously considered sufficient for the protection of the deep aquifer; however, geo-electrical surveys revealed the presence of permeable discontinuities in this layer that promote the diffusion of pollutants between the shallow aquifer (initially contaminated with leachate) and the deep aquifer.

In 1997, during a routine control from one of the existing monitoring wells around T0, values for boron, arsenic, and chlorides in groundwater higher than legislation limits were ascertained. Due to the lack of constructed barriers in T0 and T1, following to the new legal and technology standards for landfill construction in Italy, the idea of removing a possible source of groundwater contamination and replacing it with new state of the art landfill sectors was welcomed by the local authorities. In 2002, a project for the remediation of the area was approved, involving Landfill Mining (LFM) to completely remove waste form T0 and the construction of a new landfill instead, of the same surface area but significantly higher design volume. The remediation started in 2003; excavated waste were stores on a beside area until the first sectors of new T0 were ready.

Landfill mining with subsequent volumetric expansion provides a synergistic approach to overcoming the financial hurdles related to the execution of remediation projects. Several landfill sites in Italy are composed of numerous landfills, the older ones being old uncontrolled waste deposits, just a few meters deep and a potential source of contamination. LFM provides an opportunity to replace them with new landfills with the state-of-the-art technology. These new landfills constructed in the space previously occupied by old landfills exploit the pit volume and go up to and beyond 15 m above ground level. This increase in landfill volume allows the supplementary disposal of residual waste from the catchment area and the gate fees provide funding for the whole project. Moreover, this practice is currently the only solution to the unavailability of land for new landfills and is applicable in densely populated areas across Europe and other parts of the world.

2.2 History of the Project

In 2002, the Province of Rovigo approved the remediation project of the area. The remediation consisted in the LFM of the T0 and T1 areas (Fig. 2.1). Being T2 and T3 constructed as controlled landfills in compliance with applicable regulations, they therefore excluded from the remediation project. However, the landfill proximity to agricultural land necessitated a more rapid and immediate action. As a preliminary remediation measure, an embedded bentonite clay retention wall, 50 cm wide (Fig. 2.2) was constructed around the perimeter of T0 and T1, approximately 10 m deep to reach the clay layer and minimize leachate migration beyond the boundaries of the landfills. This operation began in April 2004 and was completed within the same year. Landfill mining of T0 began in 2004 with the excavation of waste deposited in the 1970s, and lasted till 2010. A storage area capable of containing a maximum of 100,000 m3 was constructed and used for the temporary storage of waste excavated from T0. A controlled landfill was constructed in compliance with Italian Legislative Decree 36/2003 on waste landfilling. The new T0 landfill sectors were then constructed and operated until 2012. The new landfill was used to

Fig. 2.2 Evidence of the embedded bentonite clay wall surrounding T0 and T1 landfill areas and separating them from the surrounding aquifer

dispose of excavated waste and residual waste after on-site mechanical biological treatment (MBT) at a nearby plant.

LFM operations on T1 commenced in 2016 and are still ongoing (2025). Results of geotechnical investigations had revealed waste buried at varying depths between −3.5 m and −4.0 m bgl, partly replacing the natural clay and silty clay layer, expected between −2 m and −6 m bgl. Excavation was performed to a depth of −4.5 m in some cells and −5 m in others to ensure complete removal of waste and potentially contaminated soil while maintaining a safe distance from the aquifer. Hydrogeological and geotechnical analyses had revealed groundwater siphoning at the bottom of the landfill cells; a significant volume of water was detected within the landfill, likely originating from the aquifers beneath it and infiltrating through discontinuities of the silty clay layer. This phenomenon could have considerably hindered the work and caused economic issues linked to the substantial production of leachate. To tackle this issue, the landfill cells were hydraulically compartmentalized prior to excavation operations to prevent water from infiltrating the site during the excavation. Steel diaphragms were installed within T1 (Figs. 2.1, 2.3, 2.4 and 2.5), dividing the landfill into sectors to facilitate the safe environmental and operational excavation of buried waste, simultaneous removal of leachate, and the construction of new landfill cells. Moreover, mechanical stability of the cells adjacent to the excavation area was guaranteed. The steel diaphragms were inserted using vibrojet technology (Figs. 2.3 and 2.4), whereby sheet piles are driven into a solid medium by vibrating equipment, thereby eliminating the need to excavate soil or waste before installation. The diaphragms are about 15–20 cm thick and were inserted to a depth of approximately 10 m, to reach the clay layer.

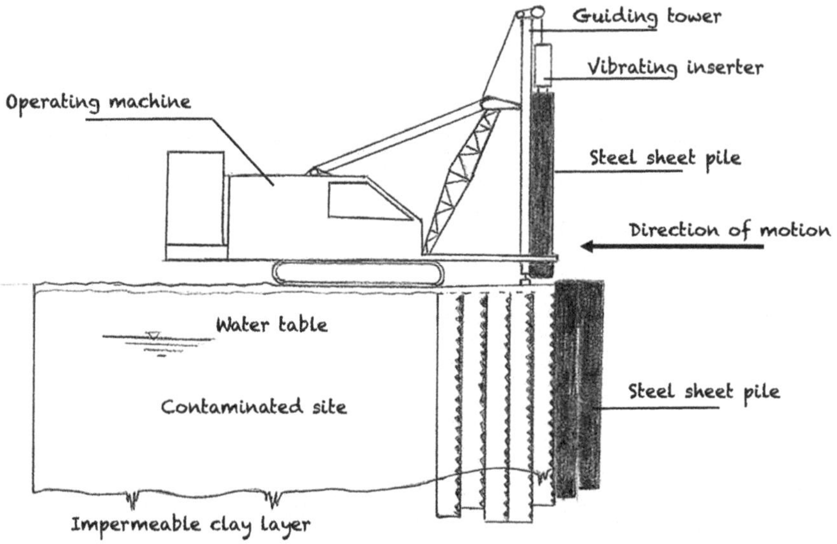

Fig. 2.3 Vibrojet technology for the installation of steel sheet piles without soil or waste excavation

Fig. 2.4 Image of machinery inserting steel sheet piles to form internal diaphragm

Fig. 2.5 Excavation works proceeding after installation of internal diaphragm

Figure 2.6a–d shows the evolution of a landfill call in T1, through the stages of landfill mining, the construction of a new landfill sector and its operational life as it is filled with waste (Ecoambiente 2012).

Fig. 2.6 Images of a landfill sector during various stages of landfill mining, new sector construction, and filling with waste. Images of a landfill sector during various stages of landfill mining, new sector construction, and filling with waste

Fig. 2.6 (continued)

The total approved volume for the new T1 is approximately 1000,000 m^3 and the volume of waste excavated from T1 is approximately 400,000 m^3, leaving an approximate 600,000 m^3 for the disposal of the new waste. With a yearly entrance of approximately 55,000 tons of new waste, the new T1 landfill is expected to serve the municipality for about 10 years.

2.3 Management of the Excavated Material and New Landfill Sectors

The old capping material (about 80 cm of soil) on the area was removed for each landfill cell before the waste excavation. These soils are classified as uncontaminated because of their position above the waste and are then stored for reuse as engineering material at the landfill itself. The waste is then excavated and either immediately deposited in new landfill sectors or in the temporary storage area for excavated waste. To efficiently remove the waste in direct contact with the bottom clay, about 5–10 cm of clay are removed along with the waste. After the waste is removed, the bottom clay at the bottom of the landfill is removed until the design excavation depth for the new landfill sectors is reached. The reclaimed clay is classified by the European Waste Code as EWC 17.05.04, that is excavated soil from contaminated sites, not containing dangerous substances. The clay is stored on an adjacent area in numbered heaps of approximately 1000 m^3 each, and then sampled for subsequent chemical analyses. In the case of compliance with the limits outlined in Italian legislation (Annex 5 to Part Four of Legislative Decree 152/2006 and subsequent amendments and additions) the soil could therefore be transferred to a storage area for future use as engineering materials in landfill construction works such as capping. In addition, the new landfill bottom is sectored and the clay sampled for analysis. If the clay at the design depth is contaminated, excavation continues and the contaminated material is removed and replaced with clean material. Then, the construction of the new barrier system for the landfill sector can begin, according to the current regulations (Ecoambiente 2020).

The waste logistics are organized so that the residual waste, pretreated at the MBT plant are immediately disposed of in new landfill sectors, together with the excavated waste. The nearby MBT plant can treat waste up to 109.200 t/year. It can also accept the unsorted fraction of the MSW collected in Rovigo province and, after conformity clearance, non-hazardous special waste from other Italian regions.

At the MBT plant, the residual waste is shredded and then sieved in a trommel screen with 100 mm diameter openings, which separates the waste into an oversieve light fraction (particle size >100 mm) and an undersieve fine fraction (particle size <100 mm). The light fraction is sent directly to the landfill. Because of its high organic matter content, the undersieve is biostabilized under aerobic conditions in a procedure similar to composting for approximately four weeks. The resulting material is soil-like with fragments of undecomposed waste and is used as daily cover at the landfill. Based on records of landfilled waste, new landfill sectors typically contain the light fraction of residual waste (New Waste Light, NWL), the Excavated Waste (EW) and the Stabilized Fine Fraction (SFF) in the mass proportions 70:20:10 respectively. The fractions of waste destined to disposal in the new landfill sectors are displayed in Fig. 2.7.

2.3.1 Waste Characterization

Samples of excavated waste for the characterization were collected in different moments at varying depths in the landfills. The procedure consisted in placing roughly 200 kg of waste on a geotextile onsite, subsequent homogenisation and further sampling using the quartering method. Approximately 50 kg were transported to the lab, where they were characterized through fractional composition and grain size distribution analysis. The waste was classified according to particle size (>100 mm, >50 mm, >20 mm, and <20 mm) and according to the following composition: *plastic, glass, metal, textile, cardboard and wood, stones, and others.*

An example of material composition of the excavated waste (sample n.1) reported as percentages by wet weight is outlined in Table 2.1. The material with particle size <20 mm termed *undersieve* mainly was composed of soil-like material resulting from degradation of organic fractions mixed with natural soil originally used as daily cover. This fraction was present with a percentage of approximately 28.5%, which is a much lower percentage than previously reported in other studies, where soil-like fractions are present with percentages above 40% (Hogland et al. 2004; López et al. 2018; Quaghebeur et al. 2013; Somani et al. 2018). This was probably due to the presence of highly compacted materials, difficult to classify, where fines were trapped and stuck and no specific treatment for separation was carried out. The other two main fractions observed are *Plastic* and *Others* (indistinguishable materials) with approximately 22% and 21%, respectively.

The characterization of waste samples destined to disposal in the new landfill sectors was carried out in the framework of a study for the assessment of the emission potential of the new landfill. A sample of the light fraction of residual waste (NWL), another sample of excavated waste (EW) and a sample of stabilized fine fraction (SFF) were considered. The fraction <50 mm was used for the tests and size reduction to less than 20 mm performed for homogenizing the sample.

The analyses of total solids (TS) and volatile solids (VS) were carried out according to international standards. Respiration index (RI_4, mg O_2/g TS) was measured as the cumulative oxygen consumption in 4 days in the Sapromat apparatus (H + P Labortechnik, Germany), according to German regulations (Anonymous 2001). The biochemical methane potential (BMP) test was set up following the Italian national standard (UNI/TS 11703:2018), the cumulative methane production in 32 days was measured.

Fig. 2.7 (**a**) Excavated waste (EW), (**b**) Light fraction of residual waste (NWL), and (**c**) Stabilized fine fraction (SFF)

Table 2.1 Composition of excavated waste and particle size distribution for sample n.1 (Tan et al. 2023)

	Excavated waste composition (kg)								
Sieve diameter (mm)	>100	%	>50	%	>20	%	<20	Weight (kg)	%w/w
Plastic	6.8	52.03	3.18	39.21	1.68	10.22		11.66	22.18
Glass		0.00	1.11	13.69	5.7	34.67		6.81	12.95
Metal	0.05	0.38	1.99	24.54	0.2	1.22		2.24	4.26
Textile	0.36	2.75	0.32	3.95	0.22	1.34		0.9	1.71
Cardboard and wood	1.2	9.18	1.06	13.07	0.45	2.74		2.71	5.16
Stones	2.2	16.83		0.00		0.00		2.2	4.18
Others	2.46	18.82	0.45	5.55	8.19	49.82		11.1	21.11
Undersieve (<20 mm)		0.00		0.00		0.00	14.95	14.95	28.44
Total	13.07	100	8.11	100	16.44	100	14.95	52.57	100

Table 2.2 Characterization of waste samples destined to disposal in the new landfill sectors. Excavated waste (EW), light fraction of residual waste (NWL), and stabilized fine fraction (SFF)

	RI_4 (mg O_2/g DM)	Biomethane potential (NL CH_4/kg VS)	Material fraction	% w/w	% in new landfill sectors
EW	2.0	2.9	Plastics	36.4	7.3
			Cardboard and paper	0.3	0.1
			Textiles	1.3	0.3
			Glass	7.1	1.4
			Metals	0.4	0.1
			Inerts (stones and ceramics)	0.7	0.1
			Others	35.7	7.1
			Undersieve (<20 mm)	18.1	3.6
			Total	100	20
NWL	25.9	76.2	Plastics	20.8	14.6
			Cardboard and paper	26.5	18.6
			Wood	2.4	1.7
			Textiles	31.0	21.7
			Glass	0.7	0.5
			Metals	0.4	0.3
			Inerts (stones and ceramics)	1.2	0.8
			Others	9.4	6.6
			Undersieve (<20 mm)	7.6	5.3
			Total	100	70
SFF	22.1	121.9			10
Total					100

The results of RI_4 and BMP tests are reported in Table 2.2, together with the fractional composition analysis of this waste sample, also considering the mass proportions previously indicated of 70:20:10 for NWL, EW, and SFF, respectively, to calculate the expected composition of deposited waste.

The values measured for respiration index and biomethane potential are consistent with the final storage quality values suggested in the Guidelines for sustainable landfills issued by the Lombardy region (Cossu et al. 2020). Much higher values are recorded for new waste especially for the fine fraction, showing that a more effective biostabilization might be considered. Concerning the NWL fraction, a significant contribution to the emission potential has to be associated to the *Others* fraction (9.4%), consisting mainly in diapers, personal care products, medical waste and food waste.

2.3.2 Landfill Simulation Experiment

A landfill simulation experiment was conducted for the assessment of the expected emissions after disposal into the new landfill sectors (Tan et al. 2024). Four plexiglass reactors, each 106 cm high with a diameter of 24 cm (Raga and Cossu 2013); the first three reactors were filled with the three individual waste fractions. A fourth reactor, simulating real landfill conditions, was filled with mixed waste (MW), containing the three waste fractions in proportions similar to those expected at the landfill (70%, 20%, and 10% for NWL, EW, and SFF, respectively. The EW and NWL were sorted to remove stones, electronic parts, glass bottles, and large metallic fractions. They were then shredded subjected to reduce the size to approximately 5 cm, to increase homogeneity. The SFF remained unaltered with a homogeneous particle size. The reactors were equipped with a leachate recirculation and collection system composed of a system of pipes, taps, and a peristaltic pump, a temperature regulating system comprised of a heating jacket used to maintain the internal temperature at 35 °C and temperature probes installed inside the reactor. The reactors were filled with water until waste fill capacity, then with an additional 5 L of water to allow recirculation. Leachate recirculation was carried out weekly and the reactors were maintained under anaerobic conditions for 280 days.

At the end of the experiment, the authors conclude that the Mixed Waste fraction, mainly composed of the light residual waste fraction, showed a faster stabilization than the latter, probably due to the action of the microbial consortium supplied by the EW. This is shown in the graph of leachate Total Organic Carbon (Fig. 2.8) as the MW curve drops faster than the NWL curve, achieving final values similar to EW and SFF waste fractions. The practice of executing landfill mining with the joint disposal of old waste with new waste in new landfill sectors may, therefore, contribute to accelerating stabilization of residual waste destined to disposal in landfill.

In a similarly set up landfill simulation experiment, MW only was placed in three column reactors with the goal of assessing the effects of alternative management strategies on the abatement of potential emissions from the new landfill sectors. Two of these reactors received intermittent air injection at different aeration rates to simulate innovative landfill management under aerated or semi-aerobic conditions, while the third reactor operated under anaerobic conditions as a control experiment.

As an example, the evolution of TOC concentration in the extracted leachate is reported in Fig. 2.9, as a function of time and of the liquid-to-solid ratio. The study

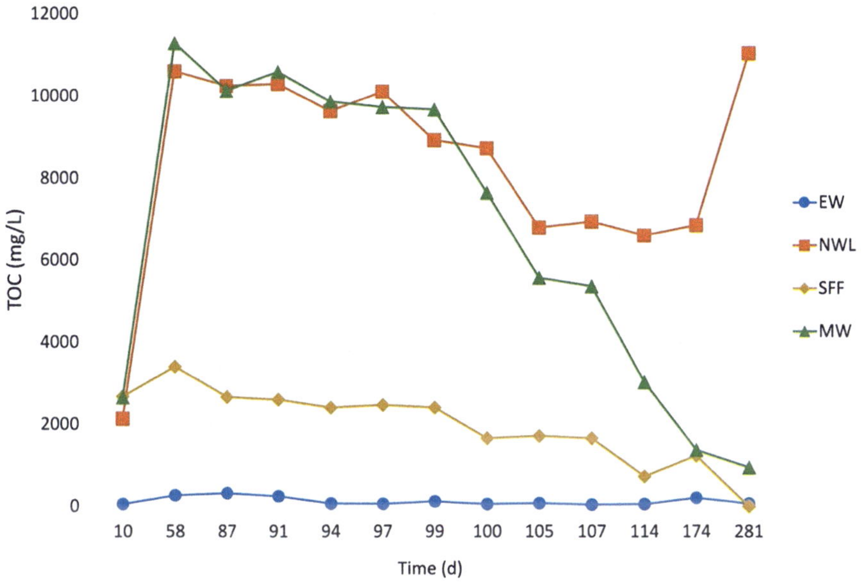

Fig. 2.8 Evolution of total organic carbon in leachate extracted from the landfill simulation reactors. *EW* excavated waste, *NWL* new waste light, *SFF* stabilized fine fraction, and *MW* mixed waste

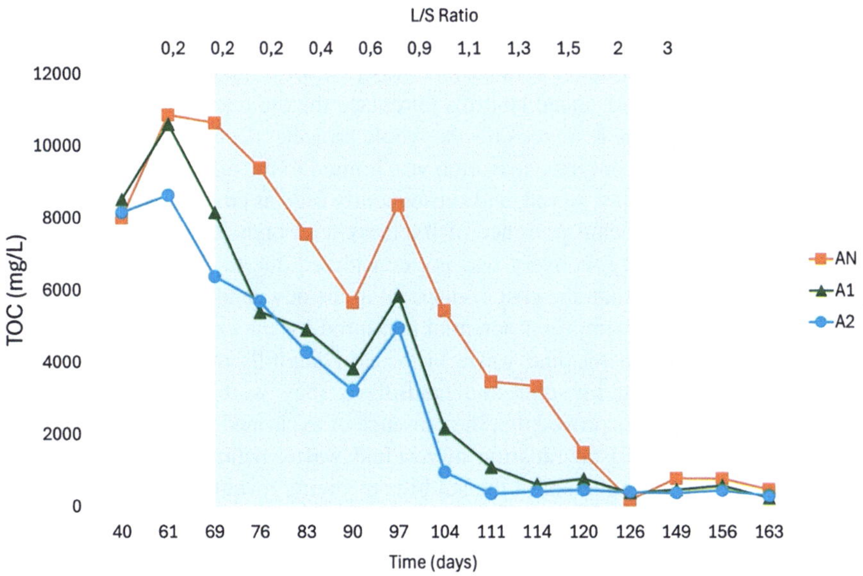

Fig. 2.9 Evolution of total organic carbon in leachate from mixed waste reactors. The blue area represents the aeration period for reactors A1 and A2

demonstrated significant differences in carbon removal pathway between anaerobic and aerated reactors. Carbon release through leachate was minimal in aerated reactors (A1 and A2), but predominant in the anaerobic reactor (AN). Generally, lower TOC concentrations were observed in leachate from the aerated reactors throughout the experiment. However, the TOC concentration for all three reactors reached similar values toward the end of the experiment as a result of the high liquid-to-solid ratios, as expected for flushing bioreactors (Cossu et al. 2003). The high liquid-to-solid ratios reached at the end of the experiment are not realistic in landfills managed with the current technology, where water inlet is on the contrary minimized during the operational and post-closure phase: in fact, despite the clear advantages of flushing toward the reduction of the long term emissions, the expected costs for leachate treatment are currently hampering its application. On the contrary, aerated landfills might have good possibilities of future development, especially in the case of semi-aerobic landfills, which involve simple technology and sustainable construction and management costs, making them suitable for low income countries as well (Hirata et al., 2012).

2.4 Conclusions

A case study of an ongoing remediation project involving the mining of an old landfill in northern Italy was presented and discussed. Groundwater contamination in the area probably due to the absence of constructed bottom barriers was the driver for the decision of completely removes the waste; however, the possibility to build new, larger, and more advanced landfills to replace the old ones has guaranteed the availability of economic resources for the whole project.

Material recovery from the excavation was limited to the non-contaminated bottom soil, mainly silty clay, stored, and subsequently used as engineering material on site. Despite the significant presence of fractions with high calorific values in the excavated waste, energy recovery was not considered due to the costs of thermal treatment, way higher than the cost of disposal in the new landfill sectors.

Due to the very low-emission potential measured for the excavated waste, placing them together with residual waste in the new landfills might seem a missed opportunity of making low-emission landfills if they were disposed of alone. However, lab-scale tests proved that the presence of excavated waste in the mixture fostered the accelerated degradation of residual waste, actually providing better conditions for the management of the landfills receiving residual waste.

Despite being designed and managed according to the current legislation and although the risk of groundwater pollution is minimized, the new landfills cannot be considered sustainable as the duration of the aftercare phase cannot be estimated as well as the related costs. This is currently an issue and the debate is in progress among the scientific community: innovative landfill models such as the bioreactor and the semi-aerobic landfills have been tested for a long time and full scale applications are available as well; however, standard landfill technologies are still far from those models.

References

Anonymous (2001) Ordinance on environmentally compatible storage of waste from human settlements and on biological waste treatment facilities, Berlin, Germany 20.02.2001

Cossu R, Raga R, Rossetti D (2003) The PAF model: an integrated approach for landfill sustainability. Waste Manag 23:37–44. https://doi.org/10.1016/S0956-053X(02)00147-2

Cossu R, Sciunnach D, Cappa S, Gallina G, Grossule V, Raga R (2020) First worldwide regulation on sustainable landfilling: guidelines of the Lombardy region (Italy). Detritus 12:114–124. https://doi.org/10.31025/2611-4135/2020.14001

Ecoambiente (2012) Relazione tecnica generale—Bonifica e recupero volumetrico della discarica Taglietto 1 (Technical report by Ecoambiente, the company managing the remediation of Taglietto 1 landfill and the design of the new landfill sectors)

Ecoambiente (2020) Relazione ultimazione vasca 5a (Technical report by Ecoambiente, the company managing the remediation of Taglietto 1 landfill, referring to the completion of sector 5a)

Hirata O, Matsufuji Y, Tachifuji A, Yanase R (2012) Waste stabilization mechanism by a recirculatory semi-aerobic landfill with the aeration system. J Mater Cycles Waste Manag 14:47–51. https://doi.org/10.1007/s10163-011-0036-7

Hogland W, Marques M, Nimmermark S (2004) Landfill mining and waste characterization: a strategy for remediation of contaminated areas. J Mater Cycles Waste Manag 6(2). https://doi.org/10.1007/s10163-003-0110-x

López CG, Küppers B, Clausen A, Pretz T (2018) Landfill mining: a case study regarding sampling, processing and characterization of excavated waste from an Austrian landfill. Detritus 2(June):29–45. https://doi.org/10.31025/2611-4135/2018.13664

Quaghebeur M, Laenen B, Geysen D, Nielsen P, Pontikes Y, Van Gerven T, Spooren J (2013) Characterization of landfilled materials: screening of the enhanced landfill mining potential. J Clean Prod 55:72–83. https://doi.org/10.1016/j.jclepro.2012.06.012

Raga R, Cossu R (2013) Bioreactor tests preliminary to landfill in situ aeration: a case study. Waste Manag 33:871–880. https://doi.org/10.1016/j.wasman.2012.11.014

Somani M, Datta M, Ramana GV, Sreekrishnan TR (2018) Investigations on fine fraction of aged municipal solid waste recovered through landfill mining: case study of three dumpsites from India. Waste Manag Res 36(8):744–755. https://doi.org/10.1177/0734242X18782393

Tan, P., Giacetti, W., Raga, R., Villani, E. (2023). Landfill remediation by landfill mining: case study of the Villadose landfill. Proceedings of Sardinia 2023, 19th international symposium on waste management and sustainable landfilling

Tan P, Raga R, Villani E, Giacetti W (2024) Landfill mining in the framework of integrated waste management: case study of an Italian landfill. Detritus 28:161–171. https://doi.org/10.31025/2611-4135/2024.19403

Chapter 3
Current Status of Landfill Mining of Old Dumpsites in India

Mohit Somani, Manoj Datta, and Manish Sharma

3.1 Introduction

India currently faces a pressing challenge in managing its legacy municipal solid waste (MSW), with approximately 10,000 ha of valuable urban land occupied by dumpsites (CPCB 2019). These unscientific landfills not only consume scarce urban space but also pose severe environmental and public health risks (Somani et al. 2018, 2020; Hölzle et al. 2022). Reducing both the volume and spatial footprint (including the towering heights) of these waste dumps has thus emerged as a national priority.

The total number of dumpsites in India is 2429, primarily those with incoming disposal of waste above 1000 tones (SBM Dash board). Several metropolitan cities are witnessing alarming vertical growth of their dumpsites, far exceeding regulatory thresholds. For instance, in Delhi, Ghazipur, Okhla, and Bhalswa dumpsites have reached heights of 62–69 m, 55–60 m, and 56–62 m (in early 2020s), respectively, which is well beyond the prescribed maximum height range of 20–40 m (Datta et al. 2021). These excessive elevations not only present significant environmental hazards but also highlight the urgent need for scientific interventions like landfill mining and reclamation to restore urban land and mitigate associated risks.

To accelerate the remediation of legacy waste, the CPCB issued a directive on January 27, 2021, under Section 5 of the Environment (Protection) Act, 1986, mandating all State Pollution Control Boards (SPCBs) and Pollution Control Committees (PCCs) to initiate and oversee landfill(bio) mining activities for legacy dumpsites.

M. Somani (✉) · M. Sharma
School of Infrastructure, Indian Institute of Technology Bhubaneswar, Khurda, Odisha, India
e-mail: mohitsomani@iitbbs.ac.in

M. Datta
Department of Civil Engineering, Indian Institute of Technology Delhi, New Delhi, India

M. Somani et al. (eds.), *Landfill Mining of Legacy MSW: Global Insights and Indian Experience*, Springer Transactions in Civil and Environmental Engineering, https://doi.org/10.1007/978-981-95-3959-8_3

This directive marked a critical step in institutionalizing landfill mining as a nationwide environmental priority.

3.2 Key Terminologies

Landfill mining has emerged as a critical waste management strategy in India, particularly in the context of urban land reclamation, environmental protection, and resource recovery. This section outlines key terminologies and processes associated with landfill mining, highlighting both international and Indian perspectives.

Legacy Waste It refers to the unsegregated municipal solid waste that has been disposed of in an unlined dumpsite (with a nominal cover or no soil cover) more than 10 years ago.

Landfill Mining (LFM) Landfill Mining is an approach whereby solid waste that has previously been landfilled is excavated and processed to recover reusable resources. It is one of the possible alternatives for reducing accumulated old waste at a dump site and reclaiming the site for other purposes or expanding landfill capacity.

Bioremediation and Biomining (Indian Context) In the global context, to any process that uses living organisms to extract metals from ores and other solid materials. Biomining refers to the excavation of old dumped waste and make windrow of legacy waste thereafter stabilization of the waste through bioremediation, i.e., exposure of all the waste to air along with use of bio-cultures, i.e., screening of the stabilized waste to recover all valuable resources (like organic fines, bricks, stones, plastics, metals, clothes, rags, etc.) followed by its sustainable management through recycling, co-processing, road making, etc. (CPCB 2019).

3.3 Progress in Dumpsite Remediation (Reclamation) via Landfill Mining (Early 2025)

According to the SBM dash board, a comprehensive inventory of dumpsites across India has been developed, providing a clear overview of the remediation status of these sites. The data, illustrated in Fig. 3.1, presents the geographical distribution of dumpsites, with a particular focus on those containing more than 1000 tons of legacy waste.

As per the current statistics, 42% of the total dumpsite area has been successfully reclaimed, while the remaining 58% is yet to be remediated (SBM dash board). Tables 3.1, 3.2, 3.3 and 3.4 provide a comprehensive quantitative overview of the scale, progress, and regional disparities in LFM initiatives across India.

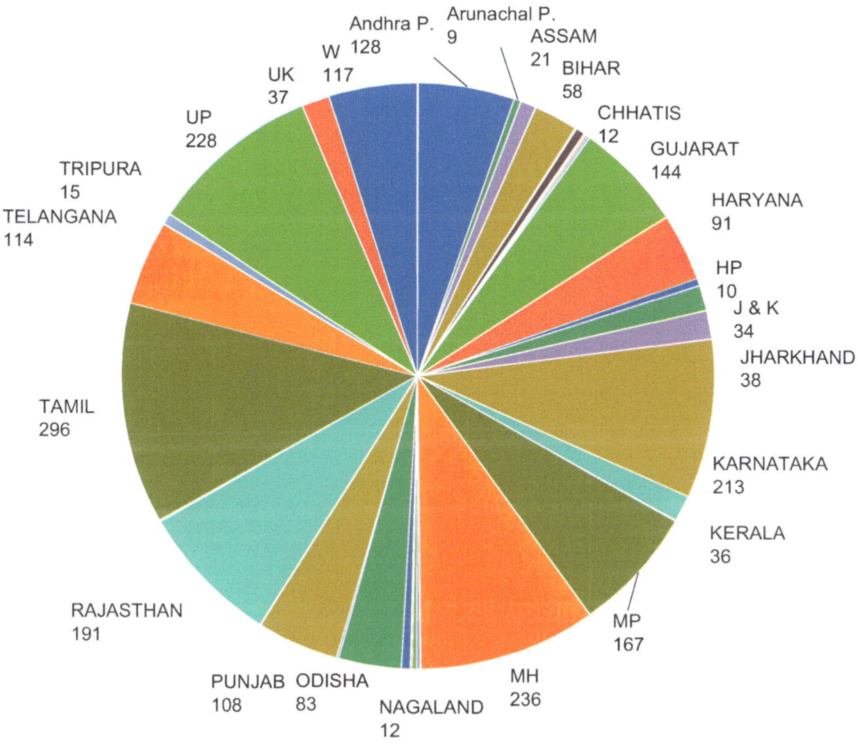

Fig. 3.1 Statewise details of dumpsites in India

Table 3.1 Status of legacy waste remediation in India (Feb 2025)

Parameter	Quantity
Total no. of dumpsites	2429
Area covered by dumpsites	14887 acre
Total area reclaimed by landfill mining	6221 acre
Total area required to be reclaimed	8666 acre
Total legacy waste present in dumpsites	2358
Total remediation of waste qty. (in lakh tons)	1163
Total legacy waste to be remediated (in lakh tons)	1193

Table 3.1 outlines the national status of legacy waste remediation as of early 2025. Of the total 2429 dumpsites occupying approximately 14,887 acres, nearly 6221 acres (42%) have been successfully reclaimed, with 8666 acres still awaiting remediation. In terms of waste quantity, out of an estimated 2358 lakh tons of legacy waste, 1163 lakh tons have been remediated, leaving about 1193 lakh tons yet to be processed. This nearly equal split between remediated and pending areas

Table 3.2 Status of dumpsite remediation through landfill mining (Feb, 2025)

Remediation status	No. of dumpsite	Legacy Waste Qty. (In Lakh Ton)	Remediated Waste Qty (In Lakh Ton)	Waste to be Remediated (In Lakh Ton)	Area (In Acre)	Area Reclaimed (In Acre)	Area to be Reclaimed (In Acre)
Remediation Completed	693	684	684 (100%)	0 (0%)	3961	3961 (100%)	0 (0%)
Remediation Ongoing	614	694	312 (45%)	381 (55%)	4894	1508.65 (31%)	3386 (69%)
Remediation Action Plan Approved	514	594	152 (26%)	442 (74%)	2925	573.25 (20%)	2352 (80%)
Remediation taken up by State/Other Funding	60	99	9 (9%)	90 (91%)	306	48.30 (16%)	258 (84%)
Balance	548	288	7 (3%)	281 (97%)	2799	129.86 (5%)	2670 (95%)

underscores the progress achieved due to policy initiatives and funding support but also emphasizes the need for sustained intervention to meet national targets.

Table 3.2 breaks down the remediation status into five categories: completed, ongoing, planned (action plan approved), funded (by state or other agencies), and untouched (balance). Out of 2429 dumpsites, remediation is completed at 693 sites (684 lakh tons of waste fully addressed and 3961 acres reclaimed). Another 614 sites are undergoing remediation, but with only 45% of the waste and 31% of the area processed so far. Furthermore, although 514 sites have approved remediation action plans, the actual execution remains limited (only 26% of waste and 20% of area remediated). Alarmingly, 548 sites are in the "balance" category, where only 3% of the waste and 5% of the land have been addressed—indicating either administrative delays or funding constraints. This table clearly illustrates that while a significant number of sites are in progress or planned for remediation, the conversion from planning to execution remains inconsistent across states.

Table 3.3 presents a state-by-state breakdown, revealing inter-state variations. States like Gujarat, Maharashtra, Tamil Nadu, and Andhra Pradesh show higher levels of reclaimed area and remediated waste. For instance, Gujarat reclaimed 873 acres out of 930 acres and remediated 216 lakh tons of waste.

Table 3.4 highlights where landfill mining has been completed, providing a cumulative snapshot of successfully remediated sites. A total of 693 dumpsites across India have been fully remediated, resulting in the reclamation of 8986 acres of land and the remediation of 684 lakh tons of waste. Notably, states like Gujarat (117 sites), Maharashtra (113), and Tamil Nadu (117) have led in terms of both the number of sites and quantity of waste remediated.

Table 3.3 Status of landfill mining in states and union territories of India (Feb, 2025)

States	Total no. of dumpsites	Area covered by dumpsites (Acre)	Area Reclaimed (Acre)	Area to be Reclaimed (Acre)	Legacy waste in dumpsites (Lakh Ton)	Remediated Waste Qty (Lakh Ton)	Waste to be Remediated (Lakh Ton)
Andaman and Nicobar Islands	1	3	3	0	1	1	0
Andhra Pradesh	128	1025	309	717	86	42	44
Arunachal Pradesh	9	49	5	44	0	0	0
Assam	21	119	13	106	25	8	17
Bihar	58	274	99	175	34	12	22
Chandigarh	2	28	20	8	13	13	0
Chhattisgarh	12	87	83	4	7	7	0
Daman Diu and Dadra Nagar Haveli	2	10	6	4	1	1	0
Delhi	3	202	0	202	203	77	126
Goa	4	15	1	14	18	1	17
Gujarat	144	930	873	57	221	216	4
Haryana	91	389	258	131	107	72	36
Himachal Pradesh	10	21	1	20	5	3	3
Jammu and Kashmir	34	126	20	105	63	2	60
Jharkhand	38	141	8	133	31	1	30
Karnataka	213	1264	69	1194	175	8	167
Kerala	36	171	57	114	13	4	9
Ladakh	1	28	0	28	1	0	1
Madhya Pradesh	167	804	410	394	59	28	31
Maharashtra	236	1457	842	614	379	191	189

(continued)

Table 3.3 (continued)

States	Total no. of dumpsites	Area covered by dumpsites (Acre)	Area Reclaimed (Acre)	Area to be Reclaimed (Acre)	Legacy waste in dumpsites (Lakh Ton)	Remediated Waste Qty (Lakh Ton)	Waste to be Remediated (Lakh Ton)
Manipur	5	11	5	6	2	1	1
Meghalaya	6	8	0	8	4	0	4
Mizoram	2	3	3	0	8	8	0
Nagaland	12	52	0	52	8	0	8
Odisha	83	416	281	135	42	17	24
Puducherry	3	19	9	10	3	3	0
Punjab	108	540	201	338	71	31	40
Rajasthan	191	1875	593	1283	108	49	58
Sikkim	2	21	0	21	3	1	1
Tamil Nadu	296	1917	980	937	191	92	99
Telangana	114	1042	382	660	158	128	29
Tripura	15	39	27	12	6	4	2
Uttar Pradesh	228	962	566	396	155	125	30
Uttarakhand	37	142	49	93	6	8	7
West Bengal	117	698	47	651	155	9	133
Total	**2429**	**14,888**	**6221**	**8666**	**2362**	**1164**	**1194**

Table 3.4 Number of dumpsites where landfill mining has been completed (Feb 2025)

States	Total no. of dumpsites	Remediated waste Qty (lakh ton)	Area covered by dumpsites (acre)	Area reclaimed (acre)
Andaman and Nicobar Islands	1	1	3	3
Andhra Pradesh	7	11	5074	5074
Arunachal Pradesh	1	0	5	5
Assam	3	1	7	7
Bihar	6	4	49	49
Chandigarh	1	5	20	20
Chhattisgarh	11	7	80	80
Daman Diu and Dadra Nagar Haveli	0	0	0	0
Delhi	0	0	0	0
Goa	1	0	1	1
Gujarat	117	200	785	785
Haryana	43	50	206	206
Himachal Pradesh	0	0	0	0
Jammu and Kashmir	0	0	0	0
Jharkhand	3	1	6	6
Karnataka	2	1	17	17
Kerala	4	2	21	21
Ladakh	0	0	0	0
Madhya Pradesh	60	8	268	268
Maharashtra	113	105	612	612
Manipur	1	1	5	5
Meghalaya	0	0	0	0
Mizoram	2	8	3	3
Nagaland	0	0	0	0
Odisha	50	8	167	167
Puducherry	1	2	8	8
Punjab	49	8	134	134
Rajasthan	23	7	187	187
Sikkim	0	0	0	0
Tamil Nadu	117	49	679	679
Telangana	6	122	281	281
Tripura	7	4	26	26
Uttar Pradesh	54	78	329	329
Uttarakhand	10	2	13	13
West Bengal	0	0	0	0
Total	**693**	**684**	**8986**	**8986**

3.4 Issues Yet to Be Addressed

Despite the measurable progress in initiating LFM projects across India, several critical aspects remain insufficiently addressed. These gaps, if left unrectified, could undermine the sustainability, safety, and long-term utility of reclaimed sites. The key issues that require urgent attention include:

3.4.1 Remediation of the Site After Landfill Mining

While the primary focus of most LFM projects is on volume reduction and material recovery, the post-mining remediation of the excavated dumpsites is often neglected or only partially planned. Key issues include:

- Residual contamination: Subsurface soils, groundwater, and the remaining fine fractions (soil-like material or SLM) often retain high levels of pollutants, especially heavy metals and organics, requiring site-specific remediation.
- Absence of capping and leachate collection: if the waste depth left at site after lfm in large, unless scientifically capped with low-permeability materials and equipped with leachate management systems, the site may continue to pose risks to the environment.

3.4.2 Inadequate Integration of Environmental Impact Assessment (EIA)

Most LFM projects currently lack a structured and comprehensive EIA framework. While EIA is mandated for new solid waste management facilities under Schedule I of the EIA Notification, 2006 (MoEF, GoI – OA 22-19/2017), there is ambiguity regarding its applicability to LFM activities involving legacy waste. As a result:

- Potential environmental risks such as air and groundwater contamination, contaminants release, noise pollution, and fugitive dust emissions are not rigorously evaluated.
- Stakeholder consultations and public participation which are the essential components of the EIA process are often bypassed, limiting transparency and community acceptance.

There is a pressing need to develop specific EIA guidelines for LFM projects that address the unique risks associated with excavating, processing, and disposing of old municipal solid waste.

3.4.3 *Post-mining Land Use*

Another major gap in the current approach is the lack of definitive planning for the after-use of reclaimed sites. Though some dumpsites have been converted into parks, bus depots while many others remain barren or underutilized. The challenges include:

- Absence of land suitability studies to assess the geotechnical and environmental feasibility for future uses such as construction, agriculture, or infrastructure development.
- Institutional ambiguity regarding the ownership, transfer, and governance of reclaimed land, especially when multiple agencies (ULBs, SPCBs, development authorities) are involved.
- Lack of guidelines from statutory bodies (e.g., CPCB, MoEF&CC) for acceptable and sustainable post-reclamation uses of previously contaminated land.

A structured framework for land use planning and zoning after mining is urgently required, with guidelines that take into account site history, contamination levels, and urban development needs.

3.4.4 *Inadequate Data on Types and Quantities of Resources Recover and Their Characterization*

Publicly accessible data on the types and quantities of resources recovered during LFM, as well as their physical, chemical, and environmental characterization, remains sparse (Somani et al. 2023). This hampers:

- Knowledge-sharing among stakeholders and researchers;
- Evidence-based policymaking;
- Benchmarking of project outcomes and scalability assessments.

There is an urgent need for systematic documentation and dissemination of material recovery data, including detailed characterizations of different waste fractions.

3.5 Conclusions

India has made significant progress in addressing its vast burden of legacy municipal solid waste through landfill mining initiatives. As of 2025, nearly half of the total dumpsite area and legacy waste quantity have been successfully remediated. States like Gujarat, Maharashtra, and Tamil Nadu exemplify effective implementation. However, the progress remains uneven and reveals critical shortcomings that threaten the long-term success of LFM efforts. Most notably, the post-remediation phase is poorly addressed, with limited efforts in residual contamination management, scientific capping, and future land use planning. Furthermore, the absence of a clear EIA framework for LFM activities leaves many environmental risks unaddressed and weakens stakeholder confidence. Similarly, the lack of structured guidelines for post-mining land use complicates land governance and restricts the effective reuse of reclaimed sites. To unlock the full potential of landfill mining in India, it is essential to establish comprehensive environmental assessment protocols, define clear land ownership and governance structures, and implement sustainable redevelopment plans for remediated areas.

References

Central Pollution Control Board (2019) Guidelines for disposal of legacy waste (Old Municipal Solid Waste). Ministry of Environment, Forest and Climate Change, Government of India

Datta M, Somani M, Ramana GV, Sreekrishnan TR (2021) Feasibility of re-using soil-like material obtained from mining of old MSW dumps as an earth-fill and as compost. Process Safety Environ Protect 147:477–487. https://doi.org/10.1016/j.psep.2020.09.051

Hölzle I, Somani M, Ramana GV, Datta M (2022) Heavy metals in soil-like material from landfills–resource or contaminants? J Clean Prod 369:133136. https://doi.org/10.1016/j.jclepro.2022.133136

Somani, M., Datta, M., Ramana, G. V., & Sreekrishnan, T. R. (2018). Investigations on fine fraction of aged municipal solid waste recovered through landfill mining: case study of three dumpsites from India. Waste Manag Res, 36(8), 744–755. doi:10.1177/0734242X18782393

Somani M, Datta M, Ramana GV, Sreekrishna TR (2020) Contaminants in soil-like material recovered by landfill mining from five old dumps in India. Process Safety Environ Protect 137:82–92. https://doi.org/10.1016/j.psep.2020.02.010

Somani M, Harbottle M, Datta M, Ramana GV, Sreekrishnan TR (2023) Identification and assessment of appropriate remediation management techniques for the recovery of soil-like material produced in landfill mining. J Environ Manag 348:119300. https://doi.org/10.1016/j.jenvman.2023.119300

Chapter 4
Evolution of Landfill Mining Practice in India and Comparison with International Practice

G. V. Ramana, Mohit Somani, and Manoj Datta

4.1 Introduction

The management of legacy municipal solid waste (MSW) has become a pressing challenge in India, where decades of uncontrolled dumping have resulted in massive dumpsites that pose significant environmental and public health risks (Somani and Shrivastava, 2021; Roy et al. 2025). Landfill mining (LFM) has emerged as a pragmatic solution to this problem, offering both waste clearance and land reclamation benefits. While the concept of landfill mining has been explored internationally since the late twentieth century, its widespread adoption in India is relatively recent, gaining substantial momentum after 2019 under regulatory and judicial directives (Somani 2025). This transition reflects a combination of factors growing urbanization, scarcity of land, environmental concerns, and political pressure which have collectively transformed landfill mining into a nationwide remediation strategy.

Indian biomining practices are distinct in their design and execution. Although the term "biomining" suggests biologically driven processes, in practice, the approach has evolved into a mechanized, excavation-centric operation. With most of the biodegradable fraction in legacy waste already decomposed, the biological stabilization component remains limited. Instead, Indian projects emphasize high-throughput excavation, mechanical screening using trommels, and rapid segregation into coarse fractions, soil-like material, and combustible residues. This model is particularly effective in densely populated metropolitan areas where space and time constraints prevent conventional stabilization methods such as windrowing.

G. V. Ramana · M. Datta
Department of Civil Engineering, Indian Institute of Technology Delhi, New Delhi, India

M. Somani (✉)
School of Infrastructure, Indian Institute of Technology Bhubaneswar, Khurda, Odisha, India
e-mail: mohitsomani@iitbbs.ac.in

M. Somani et al. (eds.), *Landfill Mining of Legacy MSW: Global Insights and Indian Experience*, Springer Transactions in Civil and Environmental Engineering, https://doi.org/10.1007/978-981-95-3959-8_4

The result is a "high-productivity, low-to-medium quality" paradigm one that prioritizes fast clearance and land recovery over refined resource recovery or environmental safeguards.

The regulatory foundation for landfill mining in India rests primarily on the Solid Waste Management (SWM) Rules of 2016 and the Central Pollution Control Board (CPCB) guidelines of 2019. These frameworks mandate scientific remediation of legacy dumps, encourage biomining wherever feasible, and provide technical guidance on excavation, stabilization, screening, and reuse of recovered fractions. While these regulations have enabled large-scale adoption, they also highlight the absence of uniform standards for contamination assessment and end-use protocols for recovered materials issues that continue to limit the sustainability of current practices.

In contrast, international landfill mining efforts, particularly the Enhanced Landfill Mining (ELFM) approach in Europe and the United States, are embedded within the principles of circular economy. Global practices emphasize multi-stage processing, advanced sorting technologies, stringent contamination testing, and integration of recovered resources into industrial applications. Compared to India's rapid clearance model, ELFM represents a resource recovery-driven framework, seeking to maximize material and energy valorization while minimizing long-term environmental risks.

This chapter examines the evolution of landfill mining in India, the regulatory and operational context shaping its practices, and the key contrasts with international approaches. By tracing these developments, this chapter highlights both the progress achieved and the challenges that remain in aligning Indian landfill mining practices with sustainable global benchmarks.

4.2 Existing Guidelines and Regulatory Frameworks

Landfill mining and legacy waste remediation in India operate under a formal regulatory umbrella provided by the Solid Waste Management (SWM) Rules, 2016, issued by the Ministry of Environment, Forest and Climate Change (MoEFCC), and the operational guidelines of the Central Pollution Control Board (CPCB), particularly the 2019 directive. These frameworks define the permissible practices and institutional responsibilities for tackling old and operational dumpsites.

4.2.1 Solid Waste Management Rules, 2016

The SWM Rules, 2016, issued by the Ministry of Environment, Forest and Climate Change (MoEFCC), formally recognized the need for scientific closure and remediation of legacy waste dumpsites. Key provisions are as follows:

Rule 15 - Duties and responsibilities of local authorities and village Panchayats of census towns and urban agglomerations.-The local authorities and Panchayats shall,—

(zj) investigate and analyze all old open dumpsites and existing operational dumpsites for their potential of bio-mining and bioremediation and whosesoever feasible, take necessary actions to bio-mine or bio-remediate the sites;

(zk) in absence of the potential of bio-mining and bioremediation of dumpsite, it shall be scientifically capped as per landfill capping norms to prevent further damage to the environment. The by-laws shall apply to every urban local body, outgrowths in urban agglomerations, Cantonment boards, Panchayat, Industrial and Institutional Townships, railways and defense establishments.

Additionally, Schedule I(j) offers technical options for the closure and rehabilitation of old dumpsites:

(i) Reduction of waste by bio-mining and waste processing followed by placement of residues in new landfills or capping as in (ii) below.
(ii) Capping with solid waste cover or solid waste cover enhanced with geomembrane to enable collection and flaring/utilization of greenhouse gases.
(iii) Capping as in (ii) above with additional measures (in alluvial and other coarse-grained soils) such as cut-off walls and extraction wells for pumping and treating contaminated ground water.
(iv) Any other method suitable for reducing environmental impact to acceptable level.

4.2.2 Central Pollution Control Board (CPCB) Guidelines 2019

The CPCB guidelines of 2019 provide a technical-operational framework for the remediation of legacy dumpsites through biomining and bioremediation. According to the guidelines, biomining involves the excavation of legacy waste followed by its stabilization and segregation into usable fractions. The key process steps include:

(a) Excavation and Windrowing: Old dumped waste is excavated and arranged into windrows—linear heaps approximately 2 m in height—allowing exposure to air for aerobic degradation.
(b) Bio-stabilization: Bio-cultures, typically consisting of composting bacterial or fungal consortia, are sprayed to promote microbial degradation of organic fractions. This step aims to reduce moisture content, odor, and leachate potential while partially decomposing residual organics.
(c) Screening and Segregation: Stabilized waste is passed through trommel screens of various sizes (e.g., >150 mm, 80–100 mm, 30–50 mm, <30 mm), separating it into recoverable fractions including:

   - Inerts and recyclables (bricks, stones, plastics, metals, rags),
   - Soil-like fines or "bio-earth" and.
   - Combustibles for RDF or co-processing.

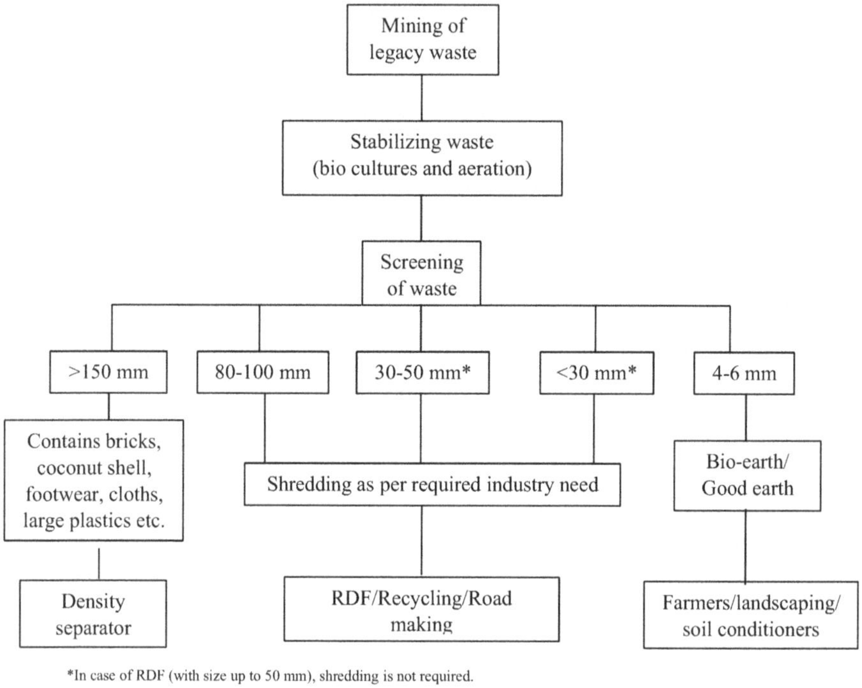

Fig. 4.1 An overview of biomining procedure in India (Central Pollution Control Board, 2019)

(d) Resource Valorization: The segregated fractions are then directed to respective end uses—recycling units, RDF plants, road construction, or soil conditioning (particularly for fractions <4 mm).

The CPCB further recommends the use of density separators to enhance material recovery efficiency and ensure better classification of heterogeneous fractions. Particularly, in cases where RDF particle size is already <50 mm, shredding is deemed unnecessary. The overview of bioremediation and bio-mining of legacy waste is shown in Fig. 4.1.

4.2.2.1 Process of Bioremediation and Biomining in India

The biomining process in India, as outlined in CPCB, (2019) guidelines, involves a structured yet flexible sequence of operations aimed at the aerobic stabilization and mechanical segregation of legacy municipal solid waste. The objective is to reduce moisture content, suppress odors, and partially degrade organic matter, thereby making the waste suitable for efficient screening and valorization.

(a) Aerobic Stabilization of Legacy Waste: The first step involves the excavation of old waste from the dumpsite using mechanical equipment such as backhoe loaders. The excavated material is then exposed to atmospheric air to initiate aerobic degradation processes. This is achieved by forming heaps (windrows) approximately 2 m in height and allowing natural aeration over time. To accelerate and enhance microbial degradation, composting bio-cultures are sprayed over the heaps. These cultures contain specific bacterial or fungal strains that assist in breaking down the remaining biodegradable fraction and help reduce the volume of waste by 30–40%. This bio-stabilization also helps in minimizing leachate formation and odor emissions, especially from partially decomposed organics and moisture-rich materials.

(b) Turning and aeration: Aerobic stabilization typically requires 3 to 4 turnings of the windrows at regular intervals. These turnings are essential to ensure uniform aeration, avoid anaerobic pockets, and maintain an optimal carbon-to-nitrogen ratio and moisture content for microbial activity. In most cases, manual labor or tractor-tillers are deployed to perform these operations, depending on the site scale and mechanization level.

(c) Types of biomining techniques: Several variants of biomining are practiced across Indian dumpsites, adapted to local conditions, equipment availability, and topography. These include:

- Windrow method (Tractor-Tiller Based): The most commonly used approach, involving linear heaps that are turned periodically using tractor-tillers or manually.
- Trench method: Excavated waste is placed into shallow trenches; aeration and microbial spraying occur within the trench. This method is space-efficient but requires dry-weather conditions.
- Cone method: Waste is piled into conical mounds, facilitating natural drainage and air exposure from multiple directions.
- Thin layer spreading: Waste is spread in thin horizontal layers (30–50 cm thick), allowing for rapid drying and microbial action due to high surface-area exposure.
- Windrow + layering hybrid: A combination of thick windrows and layered turning to balance air exposure with spatial efficiency.

(d) Timeframe: The bioremediation and bio-stabilization phase typically require 5 to 6 weeks under optimal conditions (adequate aeration, appropriate microbial inoculation, and favorable weather). During this period, a significant portion of volatile solids and moisture is reduced, enabling smoother downstream mechanical separation using trommel screens.

4.3 Evolution of Landfill Mining in India

The widespread adoption of landfill mining in India is a relatively recent phenomenon, gaining significant momentum post-2019. Biomining has been embraced as a large-scale, fast-track solution for managing legacy municipal solid waste (MSW) that has accumulated over decades in open dumpsites across the country.

The push toward landfill mining in India has been primarily driven by:

- Environmental Concerns: Unregulated dumpsites pose major health and ecological risks.
- Urban Land Scarcity: Cities such as Delhi, Mumbai, and Bengaluru face acute pressure to reclaim land for infrastructure, transport, and housing.
- Public and Political Pressure: Large dumpsites have become prominent environmental blights, attracting attention from courts, media, and political leadership.

The term Biomining, endorsed by agencies such as the Central Pollution Control Board (CPCB) and the National Green Tribunal (NGT), suggests a biological stabilization process involving the inoculation of waste with composting bio-cultures. However, field and lab-scale investigations across India (Datta et al. 2021; Somani et al. 2018, 2020, 2023) show that legacy waste typically contains up to 15% biodegradable material by weight. This indicates that most of the biodegradable content has already decomposed naturally, rendering biological treatment largely redundant in many cases. Consequently, the Indian biomining approach has evolved into a mechanized, excavation-driven operation focused on quick segregation and disposal, with minimal biological processing.

The Landfill Mining/Biomining process in India predominantly relies on backhoe excavators and rotating trommel screens as its core machinery. These equipment units form the operational backbone of the excavate-and-segregate methodology. The authors are aware, from communication with landfill operators that in large-scale dumpsite remediation projects, deployment of 20 or more excavator-trommel sets has been reported to yield productivity levels ranging from 5000 to 8000 tons per day. With each set capable of processing approximately 200 tons per 8-hour shift, the scale of material movement involved is enormous. At full operational capacity, such efforts necessitate the mobilization of over 1000 tipper trucks (assuming a 5-ton capacity per truck) every 24 h to transport segregated waste fractions away from the site (Datta 2025). In India, landfill(bio) mining approach is characterized as a high-productivity yet low-to-medium quality mining operation, primarily focused on mass excavation and coarse segregation rather than fine material recovery or high-value outputs. The process is particularly effective in extracting three major fractions from legacy MSW:

- 10–20% heavy coarse fraction (>30 mm): Primarily composed of stones, bricks, and construction & demolition (C&D) waste such as concrete debris. This material is sent to C&D processing plants to produce recycled aggregates for application in infrastructure projects.
- 10–20% lightweight sheet-like fraction (>30 mm): Consisting of soiled plastics, textiles, wood, and other combustible materials which are used in thermal power plants, waste-to-energy plants, cement manufacturing plants etc.
- 60–80% fine fraction (<30 mm): Comprising gravel and soil-like material, often moist and heterogeneous in nature, representing the bulk of the mined output. This material is used in eco-forestry, filling of low-lying areas, road works, etc.

The cost associated with the excavation and mechanical segregation phase alone is relatively modest, with estimates ranging between USD 40 and 60 per ton (equivalent to approximately INR 5 and 10 per kilogram), depending on the site-specific conditions, scale of operations, and logistics involved (Datta 2025).

In summary, India's approach can be described as "high-productivity, low-to-medium quality" landfill mining, aimed more at bulk clearance and land recovery.

The approach for biomining varies significantly based on the scale of the dumpsite, availability of operational space, and urban setting. Given the heterogeneous waste profiles, climatic zones, and infrastructural limitations, practitioners have adapted diverse strategies to achieve remediation targets efficiently. Broadly, operational strategies can be categorized based on the size of the landfill, i.e., small- and medium-sized dumpsites versus large-scale urban dumpsites.

4.3.1 Practice for Small and Medium Landfills

In the case of small- and medium-sized legacy dumpsites which are typically found in Tier-2 and Tier-3 cities or peri-urban areas, there is generally sufficient open land adjacent to or within the dumpsite boundary. This available space allows for the implementation of the classical windrow-based biomining method, as recommended by CPCB. The example of such case is shown in Fig. 4.2.

4.3.2 Practice for Large Landfills

Large legacy dumpsites, typically exceeding hundreds of thousands of metric tons and located in major metropolitan areas (e.g., Delhi, Mumbai, Bengaluru, and Chennai), present a very different set of challenges. These sites are often:

- Heavily overfilled and vertically elevated.
- Surrounded by dense urban infrastructure, leaving little to no space for traditional stabilization methods like windrowing.

Fig. 4.2 Windrows for stabilizing legacy waste at a small landfill

Fig. 4.3 Landfill mining operation at a large landfill located in Delhi

- Politically and environmentally sensitive, requiring high-speed execution under close scrutiny (Fig. 4.3).

In such contexts, the windrow stabilization method is practically unfeasible due to spatial limitations and the urgency of waste volume reduction. Consequently, the operational approach tends to bypass or minimize the bio-stabilization phase and instead focuses on:

- Direct excavation and mechanical screening: Large-scale use of backhoe excavators and trommel screens (often in 20+ unit sets) facilitates high-volume throughput (up to 8000 TPD).
- Moisture management through passive drying: Instead of structured stabilization, waste is loosely stockpiled in short-term holding areas where it is exposed to air for natural drying before screening.
- Limited use of bio-cultures: In many cases, microbial treatment is either omitted or restricted to localized patches where odor control or partial stabilization is urgently needed.
- Logistics-dominated operations: High reliance on tipper trucks and loaders for rapid removal of sorted waste fractions, often operating for long hours to meet project deadlines.

4.4 Comparison with Global Scenario

The strategic approach to LFM in India stands in sharp contrast to that adopted in developed regions such as the EU and the US. While both contexts aim to address the challenges posed by legacy waste, their motivations, operational designs, and environmental safeguards differ substantially.

4.4.1 *Indian Perspective: Rapid Clearance Model*

In India, LFM—often termed biomining—is primarily driven by:

- The urgency to clear overflowing dumpsites,
- Reclaim land for urban development, and.
- Mitigate environmental hazards under judicial and political pressure.

This results in a fast-track, low-cost operational model characterized by:

- Minimal preprocessing, limited to occasional bio-culture application,
- Rapid excavation and coarse segregation, and.
- Reuse of bulk fractions (e.g., fine soil-like materials, inert debris, and RDF) with limited contamination screening.

The focus is squarely on volume reduction and site clearance, often under aggressive project timelines and space constraints.

4.4.2 *International Practice: Enhanced Landfill Mining (ELFM)*

In contrast, the ELFM approach in the EU and US is embedded in the circular economy framework, seeking to maximize resource and energy recovery while minimizing long-term environmental liabilities. Core features include:

- Multi-stage processing involving:
  - Pre-drying,
  - Bio-stabilization,
  - Shredding and advanced sorting (magnetic, optical, ballistic, air-flow separation),
- Stringent contamination testing of fine fractions (e.g., leaching tests),
- Thorough environmental impact assessments (EIA) and life cycle assessments (LCA),
- Strong emphasis on resource valorization and sustainable reuse.
- Recovered materials—particularly metals, plastics, and high-quality RDF—are channeled into industrial applications or processed further under tight environmental controls.

4.4.3 Divergence in Reuse Practices and Safeguards

In India, recovered soil-like materials and inert fractions are often used in earthworks, road embankments, and backfilling, but without uniform contamination screening protocols. In contrast, reuse in ELFM projects is subject to:

- Regulatory clearance based on toxicity and leachability testing,
- Defined quality benchmarks, and.
- Comprehensive monitoring frameworks.

A detailed comparison between the Indian and International practice of landfill mining is tabulated in Table 4.1.

Table 4.1 Comparison between Indian and International practice of landfill mining

Category	India	International (EU/US)
Approach to landfill mining	Predominantly rapid biomining with minimal preprocessing, focused on quick clearance	Enhanced Landfill Mining (ELFM) involving Waste-to-Material (WtM) and Waste-to-Energy (WtE) strategies
Primary objective	Dumpsite clearance and land reclamation to reduce environmental and spatial burden	Recovery of valuable materials and energy; integration into circular economy principles
Excavation and processing	High-speed excavation followed by trommeling; limited or no drying and stabilization of waste	Multi-stage, technically intensive processing including drying, stabilization, and advanced segregation (magnetic, ballistic, cyclonic, air flow, and optical methods)
Material recovery focus	Emphasis on bulk reuse of recovered fractions such as SLM, C&D waste, and RDF.	Prioritized recovery of metals, plastics, and RDF; fine fractions often tested for contaminants and reused on-site or disposed under strict conditions
Geotechnical reuse	Recovered soil-like material used in filling low-lying areas, road embankments, and as earth/backfill or partial soil conditioner	Selective reuse permitted based on detailed contamination assessments; stringent quality and application standards enforced
Environmental considerations	Focus mainly on rapid site clearance; environmental and health impacts at both source and secondary disposal sites often under-assessed.	EIA, LCA, and sustainability indicators are integral to project planning and execution
Regulatory framework	Governed by NGT and CPCB guidelines; Toxicity Characteristic Leaching Procedure (TCLP) used for hazard classification	Regulated under the EU Waste Framework Directive and U.S. EPA's RCRA and relevant landfill remediation policies
Challenges faced	High moisture content reduces trommeling efficiency; inefficient segregation of mixed waste; contamination issues; policy and regulatory gaps	High cost of advanced processing technologies affects economic feasibility despite environmental advantages

4.5 Conclusions

The evolution of landfill mining in India reflects a distinctive approach driven by pressing needs for land reclamation and environmental risk mitigation. Characterized by high-productivity yet low-to-medium quality outcomes, the Indian approach prioritizes rapid clearance of dumpsites over refined resource recovery. While biomining has been operationalized at scale, its biological treatment component is often minimal, and its environmental safeguards remain underdeveloped compared to global best practices. Regulatory frameworks such as the SWM Rules (2016) and CPCB guidelines (2019) have recommend techniques to reduce moisture and odor in the waste. Guidance, yet the absence of uniform contamination assessment standards and reuse protocols for recovered fractions present ongoing challenges.

Compared to international approaches, particularly in Europe and the United States, the focus in India is more on clearing waste than on recovering high-quality materials or ensuring thorough environmental safeguards. While global practices aim for resource recovery and integration into circular economy models, the current model in India prioritizes land recovery and fast project completion. Overall, landfill mining efforts in India have proven effective for bulk waste clearance, but to ensure long-term environmental safety, there is a need for stronger regulations, better contamination screening, and more sustainable waste management strategies.

References

Central Pollution Control Board (2019) Guidelines for disposal of legacy waste (old municipal solid waste). Ministry of Environment, Forest and Climate Change, Government of India

Datta M, Somani M, Ramana GV, Sreekrishnan TR (2021) Feasibility of re-using soil-like material obtained from mining of old MSW dumps as an earth-fill and as compost. Process Saf Environ Prot 147:477–487. https://doi.org/10.1016/j.psep.2020.09.051

Datta M (2025) Insights from landfill mining of old MSW dumps in India; webinar on 27th March 2025 at the International Webinar Series on Geoenvironmental engineering, sustainability, and resiliency hosted by University of Illinois Chicago, USA

Somani M, Datta M, Ramana GV, Sreekrishnan TR (2018) Investigations on fine fraction of aged municipal solid waste recovered through landfill mining: case study of three dumpsites from India. Waste Manag Res 36(8):744–755. https://doi.org/10.1177/0734242X18782393

Somani M, Datta M, Ramana GV, Sreekrishna TR (2020) Contaminants in soil-like material recovered by landfill mining from five old dumps in India. Process Saf Environ Prot 137(2020):82–92. https://doi.org/10.1016/j.psep.2020.02.010

Somani M, Srivastava RK (2021) Regionalization of solid waste management: a proposal for divisional landfill in the state of Uttar Pradesh, India. J Hazard Toxic Radioact Waste 25(2):05020006. https://doi.org/10.1061/(ASCE)HZ.2153-5515.0000573

Somani M, Harbottle M, Datta M, Ramana GV, Sreekrishnan TR (2023) Identification and assessment of appropriate remediation management techniques for the recovery of soil-like material produced in landfill mining. J Environ Manag 348:119300. https://doi.org/10.1016/j.jenvman.2023.119300

Somani M (2025) Overcoming dumpsites rehabilitation challenges in India: landfill mining versus microbial methane oxidation systems. Indian Geotech J:1–45. https://doi.org/10.1007/s40098-025-01356-7

Roy S, Somani M, Farooq SH (2025) Microbially induced calcite precipitation for improving the Geoenvironmental performance of landfilled mined soils. Indian Geotech J:1–14. https://doi.org/10.1007/s40098-025-01346-9

Chapter 5
Case Studies of Some Landfill Mining Projects in India

Abhay Kumar Verma

5.1 Introduction

Biomining—the excavation, biological treatment, and segregation of aged waste in dumpsites—has emerged as a crucial strategy for managing India's legacy municipal solid waste. Improperly dumped waste accumulated over decades (so-called legacy waste) occupies vast tracts of land and poses severe environmental hazards. An estimated 15,000 acres of urban land (over 6000 hectares) is buried under nearly 1.6×10^8 tonnes of legacy waste across India. Open dumping is the popular MSW disposal practice in India due to budget constraints and lack of technological advancement. Open dumpsites are considered a potential threat to environment and human health since they contaminate land and water resources during migration of leachate and create air pollution in terms of release of landfill gases (LFGs) (Ghosh et al. 2022). These open dumps generate methane and leachate, causing greenhouse gas emissions, fire hazards, and groundwater pollution. In fact, decomposing landfill waste contributes roughly 3–4% of anthropogenic greenhouse gas emissions globally. Beyond climate impacts, dumpsites blight city landscapes and pose public health risks through polluted air and water and disease vectors. Recognizing these threats, policy-makers, and courts in India have pressed for systematic biomining of dumpsites to reclaim land and mitigate pollution (CPHEEO 2020; Singh 2022; Bhattacharya 2023).

Biomining follows a structured, multi-phase approach (Fig. 5.1), beginning with site assessment, survey, and planning, followed by budget approval and contractor selection. Biomining is the process in which micro-organisms are used to decompose and stabilize the organic fraction of municipal solid waste followed by

A. K. Verma (✉)
Indian Institute of Technology- BHU, Varanasi, Uttar Pradesh, India
e-mail: abhaykrverma.rs.civ17@itbhu.ac.in

M. Somani et al. (eds.), *Landfill Mining of Legacy MSW: Global Insights and Indian Experience*, Springer Transactions in Civil and Environmental Engineering, https://doi.org/10.1007/978-981-95-3959-8_5

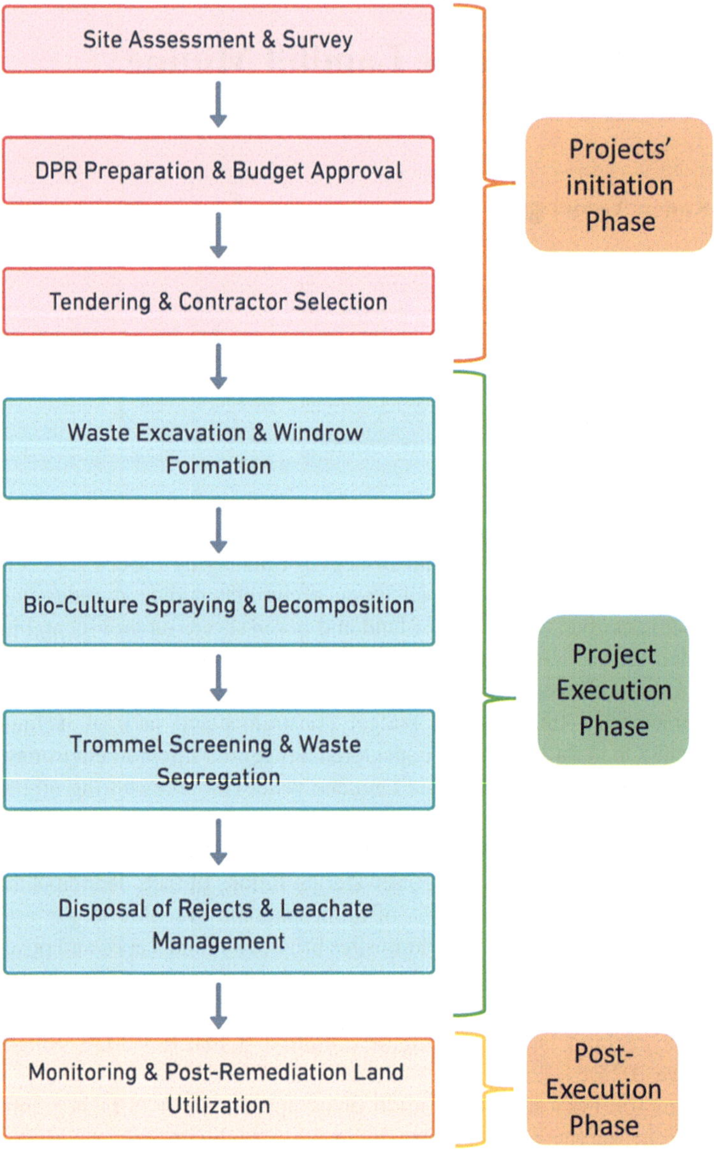

Fig. 5.1 Process flow of different Phases of a Biomining Project (Prepared by author)

excavation, treatment, stabilization, and segregation of waste (Ghosh and Kartha 2024). During the project execution phase, waste is excavated and arranged in windrows, where bio-culture spraying aids in decomposition. This is followed by trommel screening, a crucial step in segregating different fractions of waste. Rejects and leachate are then managed appropriately, ensuring environmental safety (Kumar and Kumar 2019). Finally, in the post-execution phase, the reclaimed land undergoes monitoring and remediation for sustainable utilization. This structured methodology ensures effective landfill reclamation while minimizing ecological impact.

Open dumpsites in India often appear as towering piles of mixed waste at city outskirts. Such legacy landfills emit methane (a potent greenhouse gas) and leachate, leading to fires and contamination of soil and water. Biomining is aimed at remediating these dumps to recover land and resources (Pawade et al. 2022; Bhattacharya 2023).

While biomining is fundamentally an environmental initiative, its on-ground execution in any city or Urban Local Body (ULB) requires significant legal oversight and administrative coordination. Although core technologies remain consistent, variations in site conditions allow room for innovation. This paper presents case studies from three diverse cities in Uttar Pradesh-India's most populous state—selected for their contrasting scales, geographies, and implementation models. Each case is analyzed through background, challenges, best practices, and legal compliance to aid scholarly understanding.

5.2 Legal Frameworks in Biomining

5.2.1 Evolution of Solid Waste Rules (2016 and 2024)

A strong legal framework underpins India's biomining initiatives. These rules, for the first time, explicitly addressed old dumpsites. Rule 15 (of Solid Waste Management Rules, 2016) obligates Urban Local Bodies (ULBs) to identify and remediate all old open dumpsites via biomining or bioremediation wherever feasible. If biomining is not feasible, the rules allow scientific capping of dumps with liners and vegetation to contain pollution. The SWM 2016 Rules thus made it a duty of municipalities to examine dumpsite reclamation options, prioritizing waste reduction and resource recovery over mere containment. Schedule I (Clause J) of the rules likewise lists "Reduction of waste by biomining" as the preferred approach for closure of legacy dumps, with capping as a last resort. These provisions laid the legal foundation for biomining projects across India (CPHEEO 2020; Gupta and Arora 2021; MoRTH 2023).

After 8 years of the 2016 rules, the government has moved to update the framework. In December 2024, MoEFCC released draft SWM Rules, 2024 for public consultation. The new draft rules aim to strengthen waste management through several reforms, though they largely reaffirm the earlier mandates. Notably, the 2024 draft emphasizes strict source segregation and accountability; it empowers sanitation workers to refuse collection or levy fines for unsegregated waste at source. The scope of responsibilities is widened to bulk waste generators like hotels, malls, institutions, and residential complexes, requiring them to segregate waste and compost biodegradables on-site within one year of the rules coming into force (Mohan

2024). The draft also proposes an environmental compensation (penalty) regime based on the "polluter pays" principle for non-compliance (MoEFCC 2024). By including rural areas (e.g., mandating gram panchayats to prevent burning of farm waste) and promoting circular economy approaches (such as recycling certificate credits for waste generators) (MoEFCC 2024; Mohan 2024), the SWM Rules 2024 take a more holistic view of waste management. While legacy waste remediation is not a new addition—it remains an expected duty of local bodies—the broader reforms of 2024 are poised to indirectly aid biomining by reducing fresh waste going into landfills and by incentivizing recycling efforts. The key feature of these upgradations is reflected in Fig. 5.1. The updated rules are slated to be finalized and take effect from October 2025, strengthening the legal framework for sustainable waste management in India (MoEFCC 2016, 2024; Mohan 2024).

5.2.2 NGT Directives and Regulatory Oversight

The enforcement of waste rules, especially on legacy dumps, has been spurred by the National Green Tribunal (NGT) through a series of orders. In response to widespread non-compliance with the 2016 rules, the NGT has repeatedly underscored that bioremediation and biomining of dumpsites should be the preferred option over capping for legacy dumpsites. In early 2019, the NGT directed the Central Pollution Control Board (CPCB) to issue technical guidelines for legacy waste remediation (Bhattacharya 2023). CPCB complied by April 2019, releasing detailed guidelines on biomining of old dumpsites. On the basis of these guidelines, the NGT in July 2019 explicitly declared biomining as the preferred method for legacy waste management and directed all states to "start work immediately" on clearing their dumps . Subsequent NGT orders set ambitious timelines for completing the biomining of dumpsites across India. In one such directive, the NGT recommended that legacy waste in all Class I cities be remediated by March 2023, with an extended deadline of March 2024 for million-plus population cities. For example, the Dhapa landfill in Kolkata (about 4 million tonnes of waste) was expected to be cleared by June 2024 as per an NGT-approved timeline (originally targeted for June 2024, delayed as of 2025). NGT's strict stance has been accompanied by close monitoring: it has summoned officials of lagging ULBs, imposed fines for non-compliance, and even formed oversight committees for critical cases. In its 2023–24 hearings on compliance with solid waste rules (e.g., in the *M.C. Mehta* and *Almitra Patel* cases), the tribunal expressed frustration at the slow progress, often questioning city authorities about delays (Mishra 2025). In one instance, the NGT Chair noted that "six months ago, I saw millions of tonnes of waste… where has it suddenly gone?", casting doubt on rosy reports by a municipal body (Mishra 2025). Such judicial pressure has compelled local authorities to commit to clear action plans for biomining and to

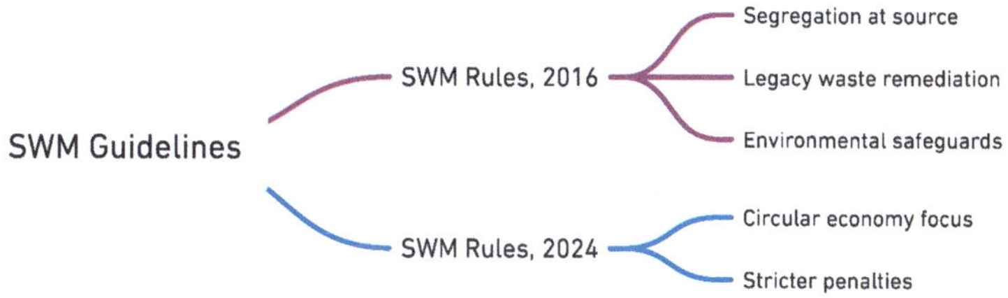

Fig. 5.2 Evolution of Solid Waste Management (SWM) Guidelines: Comparing SWM Rules of 2016 with SWM Rules of 2024 (Prepared by author)

be transparent about waste disposal claims. Overall, the NGT's directives—backed by the threat of penalties—have accelerated legacy waste remediation efforts and kept biomining on the national agenda as a legal imperative (CPHEEO 2020; Shahab and Anjum 2022; Bhattacharya 2023). The different legal and regularity frameworks including NGT, SBM, EPA, PCB and their key features which are aligned to biomining practice are shown in the mind-map below (Fig. 5.2).

5.2.3 *Institutional Roles in Biomining*

Implementing biomining projects involves a multi-tier institutional framework. Urban Local Bodies (ULBs), such as municipal corporations and councils, are at the forefront that they are responsible for planning and executing biomining of dumps within their jurisdiction, as mandated by the SWM Rules 2016. ULBs must arrange project funding (through state or central schemes), tender contracts for excavation and processing of waste, and oversee on-ground operations. At the state level, State Pollution Control Boards (SPCBs) and departments of urban development or environment provide regulatory supervision and technical support. SPCBs ensure that biomining activities (which inherently disturb large quantities of decayed waste) comply with environmental standards—for instance, controlling dust, odor, and leachate discharge during operations—and they may conduct site inspections or review data as needed. The Central Pollution Control Board (CPCB) plays a key advisory and monitoring role nationally. CPCB has issued detailed guidelines on legacy waste disposal techniques (CPHEEO 2020) and often is tasked by the NGT to compile status reports on progress in different states. In some cases, the NGT has formed committees including CPCB and MoEFCC representatives to verify local claims, reflecting the central watchdog role CPCB holds (Rai 2024). Another critical institution is the Ministry of Housing and Urban Affairs (MoHUA), which administers the Swachh Bharat Mission-Urban (SBM-U); the flagship national urban sanitation program. In its Phase 2 (SBM-U 2.0, 2021–26), the mission has put strong focus on "garbage-free cities" with a target to remediate all legacy dumpsites by 2026 (Singh 2022). MoHUA channels central funds and technical guidance to ULBs for setting up waste processing plants and biomining projects. For example, ₹217 crore was recently approved under SBM-U 2.0 specifically for remediating legacy dumpsites across over 100 municipalities (Ministry of Housing & Urban Affairs 2024). SBM has also published advisories and toolkits on landfill reclamation to disseminate best practices. Beyond government bodies, judicial institutions (primarily the NGT, as discussed) ensure accountability, while public interest groups and citizens have also influenced biomining by filing petitions and participating in oversight committees. In sum, the biomining of legacy waste in India operates under a comprehensive institutional framework: local bodies execute the work on ground, guided and funded by state/central schemes, and regulated by pollution control authorities and the judiciary. Effective coordination among these entities remains key to surmounting the challenge of India's enormous legacy waste burden (Singh 2022; PIB Delhi 2022; Thakkar et al. 2023; Rai 2024). Figure 5.3 illustrates the structured roles of key institutions in biomining, highlighting how ULBs execute projects, SPCBs regulate compliance, CPCB monitors progress, and MoHUA provides policy and financial support, ensuring coordinated waste remediation efforts (Fig. 5.4).

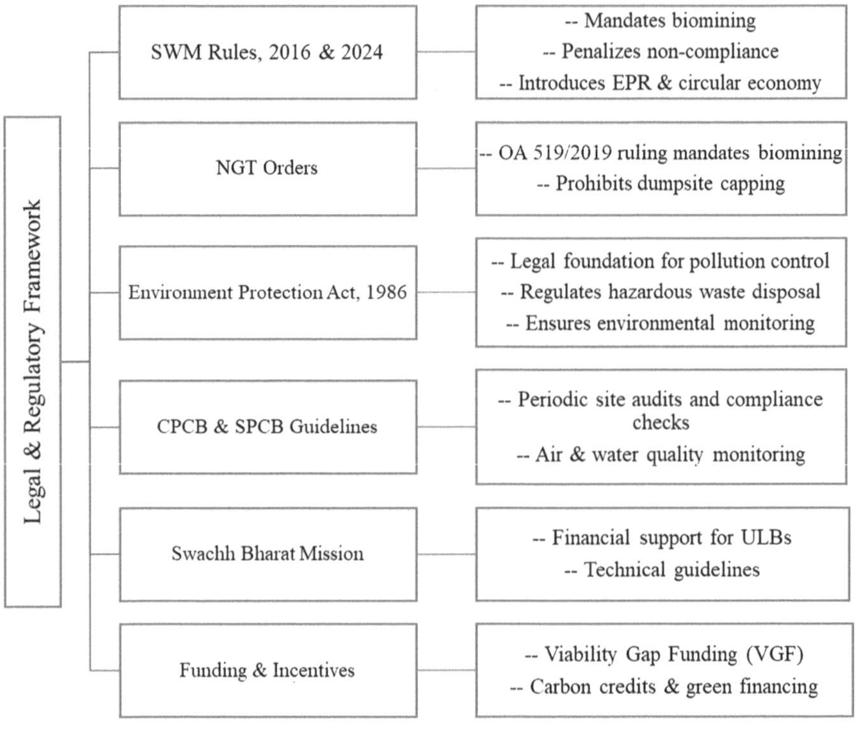

Fig. 5.3 Map shows various legal and regularity frameworks and their key features aligned to biomining practice (Prepared by author)

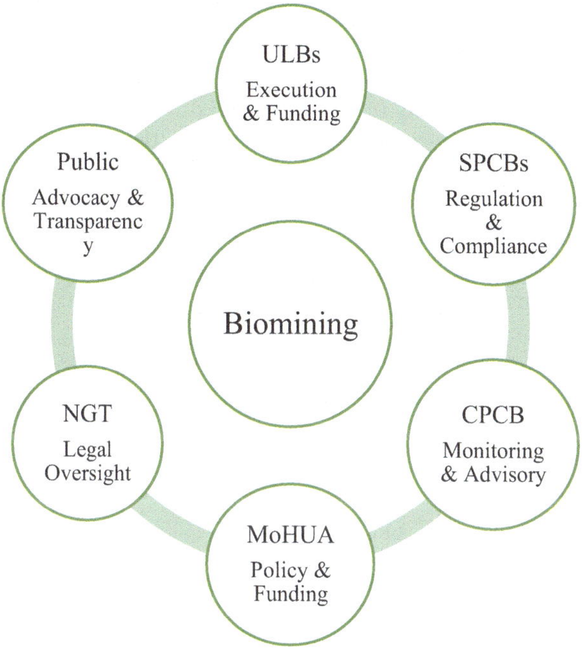

Fig. 5.4 Institutional responsibilities in Biomining; a structured framework outlining the roles of key stakeholders (Prepared in study)

5.3 Case Studies and Implementation Challenges

This section examines three such projects in Uttar Pradesh i.e. Prayagraj, Meerut, and Gorakhpur, detailing their implementation and challenges encountered. These case studies highlight common technical hurdles, such as waste segregation and leachate management, as well as institutional bottlenecks, including funding constraints and inter-agency coordination issues, which impact biomining efforts across urban centers.

5.3.1 Case Study of Meerut

5.3.1.1 Project Background

For years, Meerut, a bustling city in western Uttar Pradesh, has been grappling with massive piles of unmanaged garbage. The main dumpsite at Lohia Nagar on Hapur Road has been a growing problem, amassing hundreds of thousands of tonnes of mixed waste over the decades. By the end of 2023, the Lohia Nagar dump still held between 600,000 and 700,000 tonnes of legacy waste. NGT *observed "It is not disputed by the counsel for the municipal corporation, Meerut that as of now, around 6–7 lakh tonnes of legacy waste is still lying at Lohiya Nagar dumpsite"* (Rai 2023). Additionally, the city generates around 700 tonnes per day of fresh waste that often ends up in dumps at Gawri Village or adds to Lohia Nagar's heap (Rai 2023). The enormity of the problem led local activists (e.g. petitioners Akash Vashishth and Lokesh Khurana) to approach the NGT, which in January 2023 began strictly monitoring Meerut's legacy waste disposal progress (NGT 2023, 2024a, b). In response, the Meerut Municipal Corporation (MNN) drew up a phase-wise biomining plan and earmarked funding for the task. By early 2024, the city had allocated ₹10 crore from state funds and expected another ₹10 crore from the central government to kickstart the biomining project (Rai 2024). The initiative planned to significantly increase the dumpsite's processing capacity, jumping from 600 tonnes per day to around 4500 tonnes per day, which is almost a 7.5 times increase, by bringing in more machinery and building additional infrastructure (Down to Earth 2024). In biomining, the main product recovered is often the bioearth or good earth or the soil-like material (<8–10 mm), which can constitute greater than 60% of the legacy waste (Ghosh and Kartha 2024). This ambitious scaling-up was deemed necessary to tackle the legacy waste within a reasonable timeframe. The timeline of remediation progress at the Meerut dumpsite, including biomining and dense forest development, is illustrated in Fig. 5.5.

Fig. 5.5 Select photographs of the Biomining and Dence Forest Development in Meerut legacy waste site. (Source: MNN)

5.3.1.2 Implementation and Challenges

As Meerut's biomining project got underway, several challenges became evident. First, technical capacity had been a limiting factor. The municipal corporation had initially installed a legacy waste processing unit at Lohia Nagar that included one ballistic separator (30 ton/hour capacity) among other equipment (National Green Tribunal 2024a, b). However, processing only ~600 TPD was inadequate given the dump's size. Procuring and operating enough trommel screens, conveyors, and separators to ramp up to 4500 TPD proved challenging in terms of both cost and logistics. The city explored partnerships with public-sector undertakings—for instance, engaging GAIL (India) Ltd. and NHAI—to help establish processing or waste-to-energy facilities (Rai 2024). Utilizing support from such agencies can provide advanced technology (e.g., biomethanation or refuse-derived fuel processing) and outlets for recovered material (NHAI can use inert waste in road construction). Still, coordinating these collaborations and finalizing agreements has taken time. Meanwhile, the NGT's oversight remained stringent. The tribunal, dissatisfied with earlier slow progress, formed a high-powered committee (including CPCB and MoEFCC officials) to conduct site visits and verify Meerut's claims firsthand (Rai 2024). This level of scrutiny pressured the ULB to show tangible results quickly, but it also introduced a compliance burden—Meerut had to file frequent status reports and could face penalties if targets were not met.

On the operational front, waste segregation and treatment issues have required careful management. Decades-old waste is often heterogeneous and partially decomposed, making it difficult to segregate into usable fractions. In Meerut's case, the legacy dump contains a mix of decayed organic matter (now mostly soil-like fines), plastics, glass, metals, and a significant quantity of construction and demolition (C&D) debris given urban growth. Handling this mix necessitates multiple processing stages: excavation with excavators, coarse separation of large C&D debris, trommel screening for finer fractions, and further separation of combustibles versus inert materials. The corporation has reported efforts to augment these processes, e.g., adding more trommel screens and mechanical separators to increase throughput. However, scaling up machinery brought challenges of its own: ensuring uninterrupted power supply, machine maintenance, and trained operators for the new equipment. Another technical challenge has been leachate and odor control. The Lohia Nagar dump, like many old landfills, has pools of leachate (toxic liquid from waste) at its base. Excavation can release this leachate and produce foul odors. In the initial phases, Meerut had to deploy measures such as spraying bio-deodorizer or bio-culture solutions on the waste to suppress odors and accelerate microbial degradation (CPCB 2019). Managing leachate runoff especially during rains required creating drains or sumps to collect and treat the leachate, to prevent contamination of nearby soil and water. These environmental safety steps, while necessary, sometimes slowed down the digging and screening work. As part of its broader waste management efforts, the Meerut Municipal Corporation (MMC) also undertook the remediation of the Gawndi Jalalludin Dumpsite, covering 45 acres of legacy waste. After successfully completing the biomining and reclaiming the landfill, the area was transformed into the "Praan Vayu Upwan Centre," where they planted 25,000 trees using the Miyawaki method of afforestation, known for creating dense native forests with high biodiversity and carbon sequestration potential in limited spaces, significantly improving air quality and urban ecology (Saliba et al. 2024).

5.3.1.3 Financial and Institutional Barriers

Financing the biomining project has been a pivotal concern in Meerut. The total legacy waste (≈0.65 million tonnes) would cost on the order of tens of crores of rupees to fully remediate. The municipal budget alone was insufficient, leading to reliance on state and central grants. While ₹20 crore was lined up from government sources (Rai 2024), any shortfall could jeopardize the pace of work. In this regard, securing the additional ₹10 crore from the central SBM-U funds was critical—delays in fund disbursement from higher levels at times slowed procurement of equipment. The municipality also explored innovative financing, such as involving corporate social responsibility (CSR) funds of public companies (the mention of GAIL and other PSUs hints at possibly leveraging CSR or technical partnerships) (Rai 2024). Bureaucratic processes for such collaborations, however, can be protracted, causing further delays. Institutionally, Meerut's experience highlights the coordination challenge between agencies. The city needed clearances or support from the state urban development department, guidance from SPCB on environmental compliance, and the cooperation of contractors and technology providers—all under the watchful eye of the NGT. Initially, a lack of adequate project management capacity in the ULB contributed to slow progress (prompting the NGT's summons of the Municipal Commissioner) (Rai 2024). To overcome this, the city strengthened its project monitoring, often with direct supervision by senior officials and more frequent progress reviews. By early 2024, Meerut reported a more robust plan and some improvements on ground, but the case underscores how large-scale biomining projects require significant up-front investment and inter-agency coordination, which can be difficult for resource-constrained municipalities. Meerut's ongoing efforts—if successful in ramping up processing to thousands of tonnes per day—could become a model for other big cities with massive legacy waste, but the hurdles it faced are instructive for policymakers and practitioners alike.

5.3.2 Case Study of Gorakhpur

5.3.2.1 Project Background

Gorakhpur, a medium-sized city in eastern Uttar Pradesh, faced a significant challenge with legacy waste at its Ekla Bandha dumpsite, where approximately 196,000 metric tonnes of waste had piled up by 2022, the accumulated waste mound had reached about 196,000 metric tonnes (approx. 0.20 million tonnes). The dumpsite's proximity (within 50 m) to the river meant a high risk of leachate contaminating the water, and the rotting waste posed fire and health hazards for nearby communities. Recognizing these threats, the Gorakhpur Municipal Corporation launched a concerted biomining project in May 2022 to remediate Ekla Bandha. The initiative was notably led by an enthusiastic city leadership—the Municipal Commissioner himself (Gaurav S. Sogarwal, IAS) took personal interest, later documenting the project's outcomes and links to climate action in a case study. The objectives were not only to clear the waste, but also to transform the site into a beneficial public space and explore innovative financing like green credits. By mid-2023, Gorakhpur's biomining operation had achieved 100% remediation of the Ekla Bandha dumpsite, an impressive accomplishment within roughly a year. Approximately 1.96 lakh tonnes of legacy waste were excavated, treated, and disposed of through various resource recovery channels, completely clearing the garbage hill. The cleaned site was then converted into a "green zone" with extensive plantation. In total, about 50,000 tree saplings were planted on the reclaimed land, creating an urban forest and carbon sink for the city's benefit (Sogarwal 2025). The case study reveals that reclamation of dumpsites through biomining enables municipalities to recover land for beneficial use such as developing green spaces, and recover valuable materials for recycling and reuse (Mohan and Joseph 2020).

5.3.2.2 Implementation and Outcomes

Gorakhpur's approach to biomining was methodical and environmentally conscious. The process began with careful planning: baseline studies of the waste composition and potential environmental impact were conducted (Zhou et al. 2014). As discussed in Meetrut similar biomining approach has been implemented at the Gorakhpur dumpsite, as shown in Fig. 5.6, where waste remediation has led to successful site restoration and afforestation. Given the dump's hazardous status (emitting smoke and leachate), the corporation prioritized safety measures at each step. During excavation, legacy waste was systematically dug out using JCB loaders and excavators. Workers and machines operated in a phased manner to avoid destabilizing the heap all at once, thereby minimizing sudden gas releases or landslides of waste. The excavated material was laid out in windrows and sprayed with bioculture inoculums to accelerate decomposition of organic matter and suppress foul odors (CPCB 2019; Pawade et al. 2022). Next, the waste was passed through trommel screens and other segregation units to separate it into different fractions. By the end of the project, Gorakhpur was able to recover multiple useful fractions from the dump. Roughly 80% of the mined waste (by weight) turned out to be soil-like material—essentially stabilized organic matter and fine inert soil—totaling about 1,56,800 MT. This recovered soil was utilized for land reclamation (filling low-lying areas) and for landscaping and agricultural use after quality testing. Another portion, about 10% of the waste ($\approx$19,600 MT), was composed of bricks, stones, and other C&D waste; this was sent to a city C&D recycling plant at Mahesra for use in making aggregates or other building material. The combustible fraction—mainly

Fig. 5.6 Stepwise progress of Biomining at Gorakhpur Dumpsite, showcasing the transformation from initial waste accumulation to remediation and eventual plantation

plastics, rags, leather and such—accounted for roughly 8% (≈15,680 MT) and was processed into Refuse-Derived Fuel (RDF). This RDF was transported and co-processed in cement kilns (cement manufacturers readily accept RDF as a substitute fuel). Finally, the remaining 2% (≈3920 MT) of the legacy waste, including recyclable plastics, glass and metal scraps, was recovered and sent to recycling facilities. The sale of these recyclables even generated a modest revenue (reported as about ₹50,000) for the city. Through such comprehensive segregation, Gorakhpur achieved near-total material recovery, drastically reducing the need for residual landfilling. The project not only reclaimed around 5 acres of land at Ekla Bandha but also yielded environmental benefits: city officials noted significant reductions in methane emissions and improvement in local air quality indicators after the dump's removal (the city's PM_{10} levels improved, contributing to Gorakhpur securing a high rank in a national air quality survey). The culmination of the project was the development of an urban forest (Fig. 5.6) on the cleared site—50,000 trees (including Neem, Peepal, and other native species) were planted, which are expected to sequester carbon and provide a green lung for the city in coming years. Gorakhpur's successful biomining thus turned a public health nuisance into an environmental asset (Sogarwal 2025).

5.3.2.3 Challenges and Innovative Practices

Despite its success, the Gorakhpur biomining project faced challenges typical to legacy waste remediation. One major challenge was managing environmental risks during excavation. With the dump so close to the Rapti River, there was acute concern about leachate runoff entering the water. The project team mitigated this by constructing temporary embankments and drains and treating the leachate on-site (using lime and bio-enzymes) before discharge. Regular monitoring of groundwater around the site was instituted to detect any contamination early. Fire control was another concern—digging into a dump can expose methane pockets and cause fires. Gorakhpur tackled this by phasing the work and maintaining readiness with water tankers. In fact, by removing the bulk of decomposing waste, the project greatly reduced the spontaneous fire incidents that had plagued the area before. Another challenge was gaining public trust and participation. Initially, local residents were skeptical about the excavation, fearing increased odor or pests. The municipal team conducted outreach, explaining that short-term inconveniences would lead to long-term relief. As work progressed and the community saw the dump shrinking, support grew. The project's quick timeline (one year) also helped maintain public patience. On the operational side, Gorakhpur benefited from adopting a public–private partnership model. It engaged experienced waste management agencies for the biomining work and implemented a rigorous monitoring regimen. Use of GPS trackers on waste transport trucks and daily logging of processed quantities improved transparency. The involvement of external technical experts (including a collaboration with researchers on quantifying emissions reductions) added credibility. One innovative aspect was Gorakhpur's linkage of the biomining project with the concept of Green Credits. The Green Credit Program, which was first introduced in 2023, assigns tradable credits for environmental actions like pollution reduction, waste management, and afforestation (administered by the Indian Council of Forestry Research and Education). Gorakhpur positioned its dumpsite remediation cum afforestation project to qualify under this scheme—essentially aiming to earn tradable credits for the waste cleared and trees planted. While still a nascent idea, this could allow the city to monetize its environmental benefits by selling green credits to industries seeking to meet sustainability targets. It exemplifies how cities can combine climate finance mechanisms with solid waste management. In summary, Gorakhpur's case shows that strong local leadership, community engagement, and integrated planning (waste processing + land reuse + climate incentives) can overcome the usual barriers in biomining projects. The city's experience stands as a best-practice model for medium-sized cities dealing with legacy waste.

5.3.3 *Case Study of Prayagraj*

5.3.3.1 Project Background

Prayagraj, once known as Allahabad, is a historic city in Uttar Pradesh that dealt with one of the biggest legacy waste problems in the state. The primary dumpsite at Naini, also called Baswar, had built up about 1.4 million tonnes (14 lakh tonnes) of municipal waste over many years (Mishra 2025). This legacy waste attracted heightened scrutiny in recent years due to Prayagraj's preparation for major events like the Maha Kumbh Mela (a mass religious gathering). Between 2021 and 2022, the Prayagraj Municipal Corporation started a large-scale biomining and landfill reclamation project at Naini. The goal was to handle all 1.4 million tonnes of waste quickly, in time for the 2025 Maha Kumbh (initially aiming for completion by 2024, in time for the 2025 Maha Kumbh) (Mishra 2025). To execute this, PMC engaged multiple contractors for different facets of the work. Three biomining firms were hired to carry out simultaneous waste excavation, screening and processing on different sections of the dumpsite (DoEFCC 2023). Additionally, a logistics firm was brought on to handle the transportation and further utilization of the recovered refuse-derived fuel. Given the project's complexity and importance, Technical support was provided by external agencies (through a development agency) to advise and help implement a sustainable business model for the biomining effort (DoEFCC 2023). A notable feature of Prayagraj's plan was an integrated "waste to resource" model leveraging Extended Producer Responsibility (EPR). The city entered a tripartite Memorandum of Understanding with a sustainability firm and a cement manufacturer to channel the non-recyclable fractions of legacy waste (like low-grade plastics and other combustibles) for co-processing in cement kilns (DoEFCC 2023). Under this arrangement, refuse-derived fuel extracted from Prayagraj's legacy waste would be transported to Dalmia's cement plant in a neighboring state (Bihar) for use as alternative fuel, and in return the cement company would provide EPR certificates/credits that help plastic producers meet their recycling obligations (DoEFCC 2023). This innovative financing model essentially meant that brands (Producers, Importers, and Brand-Owners under the Plastic Waste Rules) would fund the cost of transporting and processing the waste in exchange for the resulting environmental credits. With strong backing from the Swachh Bharat Mission and state authorities, Prayagraj's biomining project was set in motion as a flagship initiative (DoEFCC 2023). To comprehensively understand the biomining project in Prayagraj, the structured process flow is illustrated in Fig. 5.7, outlining key steps from project planning to waste recovery, land reclamation, and long-term monitoring.

The Prayagraj biomining project has been as challenging as it is ambitious. Project management and monitoring have proven complex due to the multiple stakeholders and contractors involved. Coordinating three different biomining agencies working in parallel required robust oversight to ensure uniform standards and avoid inefficiencies or gaps. To address this, the PMC established a

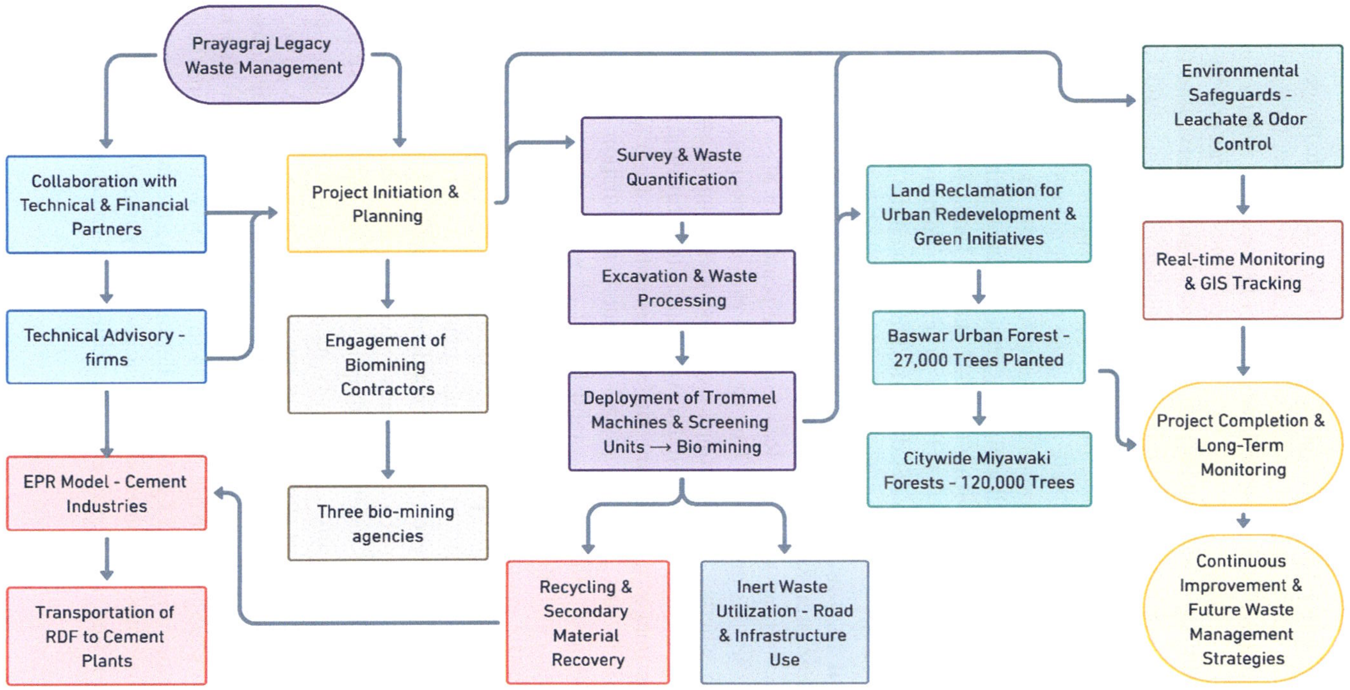

Fig. 5.7 Prayagraj legacy waste management process—a structured approach integrating biomining, waste recovery, environmental safeguards, and urban redevelopment to ensure sustainable landfill remediation and resource utilization (Prepared in study) Implementation and Challenges

centralized project monitoring unit and enlisted the help of technical experts to supervise operations (DoEFCC 2023). Still, synchronizing the workflow (excavation, on-site processing, removal of outputs) was a daily challenge, especially as each contractor brought its own equipment and workforce. One early hurdle was the quantification of waste and planning of processing capacity. Initial estimates pegged the legacy waste at ~1.4 million tonnes, but accurately mapping the spread and depth of the dumpsite required advanced surveying (including drone surveys and borehole sampling). Any underestimation could leave contractors short of capacity; any overestimation could mean wasted resources. The project planners opted for a slightly over-capacity approach—bringing in extra trommel machines to collectively achieve an output of several thousand tonnes per day. Even so, processing such a large volume in the short timeline required was daunting. As seen in similar big-city projects, factors like weather considerably affected productivity. During the 2022 and 2023 monsoons, work slowed or halted due to rains—wet waste is harder to screen, and heavy machinery movement on muddy ground becomes difficult. This slowed the overall progress and put pressure on dry-season operations to make up for lost time. An affidavit to NGT in April 2023 revealed that only about 20% (0.28 million tonnes) of the waste had been processed by Feb 2023, despite the project being almost halfway through its time schedule. Reaching the remaining 80% in the available time posed a serious challenge. This shortfall raised concerns among oversight bodies about whether the deadlines would be met (Bhattacharya 2023). The phased progress of biomining at Baswar dumpsite, Prayagraj, is depicted in Fig. 5.8. The images illustrate the site's transformation from a heavily waste-laden area (February 2022) to a well-structured remediation and processing facility (December 2024). The establishment of waste sorting units, material recovery zones, and afforestation areas demonstrates the city's commitment to sustainable waste management and land reclamation. The logistics of resource recovery in Prayagraj also presented challenges. A core idea was to send the combustible refuse (RDF) to the cement plant and to channel recyclables into recycling markets. However, moving hundreds of thousands of tonnes of material out of the city required a continuous convoy of trucks. Managing this transport and preventing any secondary dumping or spillage during transit required stringent operational control. The involvement of an experienced logistics firm helped create a dedicated supply chain for RDF (DoEFCC 2023). Still, documentation and tracking of these waste shipments had to be meticulous—any lapse could result in environmental non-compliance. The project also had to build interim storage and baling facilities for the RDF bales before they were shipped, to streamline loading. Meanwhile, managing the inert fraction (which typically constitutes 50% or more of legacy waste by mass) was another issue: PMC had to find or create sites to utilize this material (e.g., for local

Fig. 5.8 Transformation of Baswar site, Prayagraj–The site maps illustrate the evolution from a legacy waste-covered area (February 2022) to a structured waste processing and reclamation facility (December 2024) post-bioremediation. (Source: PMC)

construction, road sub-base, or for filling low-lying areas). Ensuring that the inert material didn't just end up creating another dumpsite is a key consideration (Bhattacharya 2023; DoEFCC 2023; Sogarwal 2025).

5.3.3.2 Administrative and Transparency Issues

Prayagraj's biomining effort came under intense scrutiny by the NGT, especially in late 2024, highlighting some institutional challenges. In an NGT hearing (January 2025, in the *M.C. Mehta* case), the tribunal questioned the "sudden disappearance" of 1.4 million tonnes of legacy waste that the city claimed to have achieved. The Municipal Corporation had reported that it disposed of all 14 lakh tonnes within a span of about 6 months—an assertion that was met with skepticism. The tribunal noted that as recently as mid-2024, huge piles of waste were visible, and it pressed the corporation to explain what methods were used to dispose such a massive quantity so rapidly. When the corporation's lawyer responded that the waste was treated and sent to cement plants, the NGT bench asked for specifics—"Which cement companies? What are their names?" (Mishra 2025)—details the officials could not readily provide. This pointed to a transparency and record-keeping gap in the project. It raised the possibility that either the data was being misreported or that the waste might have been hurriedly relocated (rather than truly processed) in order to showcase a clean site for the upcoming Kumbh Mela. The NGT's pointed questions indicate a concern that paper compliance might not match reality, a risk in any large project under political pressure. The tribunal has since demanded detailed affidavits from the PMC, seeking clarification and proof (such as weighbridge records, agreements with cement plants, etc.) of how the legacy waste was managed (Mishra 2025). The Prayagraj Municipal Corporation (PMC) stated that it is adhering to waste management guidelines under Swachh Bharat Mission 2.0 and has made significant progress in biomining and land reclamation. PMC emphasized scientific waste processing, real-time monitoring, and afforestation efforts, including a Miyawaki forest with 27,000 trees, while acknowledging challenges in execution and coordination but reaffirming its commitment to sustainable remediation (PMC 2024). This episode underscores the challenge of institutional accountability. Under event-driven deadlines, local authorities might resort to quick fixes (like simply flattening the garbage mound or shifting waste elsewhere) to show compliance, which defeats the purpose of sustainable waste management. Prayagraj's case thus highlights the need for rigorous monitoring by independent authorities to ensure that biomining projects achieve genuine waste processing and resource recovery, rather than just cosmetic outcomes (Mishra 2025). Figure 5.9 illustrates the phased biomining process at the Prayagraj dumpsite, showing waste excavation, screening, and site redevelopment from February 2022 to May 2024. The images highlight the structured approach taken by PMC for systematic remediation and land reclamation.

Despite these hurdles, Prayagraj's initiative has introduced some pioneering practices. The EPR-driven financing model is one: by leveraging the Plastic Waste Management Rules 2022 that mandate producers to manage non-recyclable plastic, the city tapped into private sector funds to handle legacy plastics (MoEFCC 2022; DoEFCC 2023). This model offers a "win-win" solution: the city gets financial support to move out its low-value waste, cement plants get fuel, and consumer

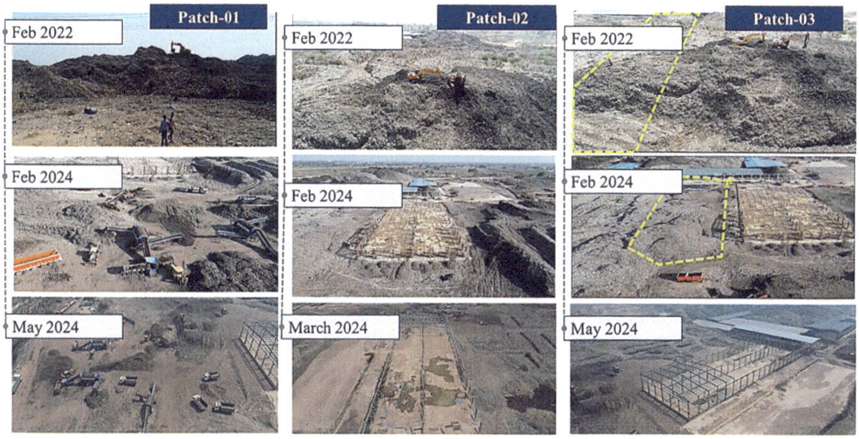

Fig. 5.9 Stepwise Progress of Biomining at Prayagraj Dumpsite—The images illustrate the phased remediation of different patches from legacy waste accumulation (February 2022) to site preparation and infrastructure development (May 2024). (Source: PMC)

goods companies earn EPR compliance credits for disposing legacy plastics (DoEFCC 2023). If executed properly, such partnerships could substantially offset the costs of biomining large dumps. Additionally, Prayagraj integrated its biomining with broader city initiatives under the Swachh Bharat Mission and Smart City program, such as deploying smart monitoring tools. There were plans to use drones and GIS mapping at the Naini site to monitor progress and detect any fresh dumping (to prevent re-accumulation). The city also explored establishing a real-time dashboard to track the tonnage processed by each contractor daily—a practice that, if implemented, can improve transparency and allow quicker troubleshooting of bottlenecks. In summary, Prayagraj's case study reveals the scale at which biomining is being attempted and the multifaceted challenges it entails: massive coordination efforts, the importance of accurate data and honest reporting, and the critical role of oversight. It also shows that innovative financial and technical solutions are being tried to make large-scale biomining feasible. The ultimate success of Prayagraj's project will depend on sustained commitment and truthful adherence to environmental objectives beyond the immediate political or event-related motives.

Given, Table 5.1 provides a comparative overview of key biomining parameters across three cities in Uttar Pradesh. It captures waste quantities, project timelines, and institutional frameworks, offering insight into how different Urban Local Bodies approached legacy waste remediation.

Table 5.1 Key Comparison Table: Biomining Initiatives in Meerut, Prayagraj, and Gorakhpur

Parameter	Meerut	Gorakhpur	Prayagraj
Dumpsite location	Lohia Nagar, Gawri Village	Ekla Bandha (Gulariha Block)	Naini (Baswar)
Estimated legacy waste	600,000–700,000 tonnes	196,000 MT (est. up to 5 lakh MT)	1.4 million tonnes (14 lakh tonnes)
Project initiation	2023	2023	2021–2022
Key implementing agencies	Meerut Municipal Corporation (MMC)	Gorakhpur Municipal Corporation (GMC)	Prayagraj Municipal Corporation (PMC)
Contractors engaged	Biomining agencies appointed by MMC	Biomining agencies appointed by GMC	Three biomining industries (ref: (DoEFCC 2023))
Waste processing capacity	Initially 600 TPD, scaled to 4500 TPD	1000–1500 TPD	Several thousand TPD
Technological support	NHAI, GAIL, RDF processing for cement industries	State and central collaboration	Advisory agencies support in RDF co-processing in cement kilns
RDF utilization	Supplied to cement plants	Exploring industrial partnerships	Tripartite MoU with cement and procurement agencies
Environmental measures	Leachate control, odor management, green buffer zones	Leachate management, odor control	Leachate treatment, GIS tracking, bio-culture spraying
Monitoring & transparency	GIS-based tracking, real-time progress reports	Periodic audits, manual inspections	Drone surveys, digital dashboards for monitoring
Reclaimed land use	Green redevelopment, Miyawaki afforestation, Praan Vayu Upwan	Proposed parks and infrastructure use	Green spaces, land reclamation, city expansion
Funding & Financial Model	~₹10 crore from state funds, additional ~₹10 crore from SBM-U	State grants and municipal budget	~₹20–30 crore from state and central grants, EPR-based financing
Challenges faced	Waste quantification issues, logistics, seasonal disruptions	Limited processing capacity, need for additional funds	Coordination of multiple contractors, transportation of RDF
Key successes	Rapid capacity scaling, industrial collaboration	Steady progress with environmental safeguards	EPR financing model, RDF-to-cement industry integration

5.4 Conclusion

This chapter has demonstrated that biomining is not just a technical intervention but a multi-dimensional approach requiring legal mandates, administrative commitment, and strategic innovation. Biomining is a simple, innovative and environmentally acceptable measure to remediate the open dumpsites for achieving zero emission of landfill gases, leachate and reclaiming the land for reuse. By examining three case studies from Meerut, Prayagraj, and Gorakhpur—each with varying waste scales, institutional capacities, and financial models—we reveal how policy frameworks, such as the Solid Waste Management Rules (2016, 2024) and NGT directives, shape implementation outcomes. Cities that embedded transparency mechanisms, stakeholder participation, and adaptive technologies—like Prayagraj's EPR-financed RDF model or Gorakhpur's green credit integration—achieved more resilient results.

While operational challenges such as waste heterogeneity, seasonal disruptions, and equipment constraints persist, this study underscores that strong governance, inter-agency coordination, and circular economy thinking can overcome structural limitations. The findings offer a scalable roadmap for other Indian cities, positioning biomining not merely as a waste disposal method but as a transformative process for environmental recovery, institutional innovation, and urban regeneration.

Declaration The author is associated with a State-level Swachh Bharat Mission project in Uttar Pradesh, India. This study is based on publicly available data and additional operational inputs obtained through author's engagement with Urban Local Bodies. While care was taken to verify information, minor variations may arise due to differences in reporting practices and interpretations across administrative sources. AI tools were ethically used for structuring and refining content. No conflict of interest is declared. For queries, the author can be contacted via the provided email.

Funding Disclosure Declaration The author acknowledges that this research did not receive any specific grant from funding agencies in the public, commercial, or not-for-profit sectors.

References

Bhattacharya S (2023) Kolkata attempts to eliminate "legacy waste" in landfills through biomining. Mongabay India

Central Pollution Control Board (CPCB) (2019) Guidelines for disposal of legacy waste (old municipal solid waste). Government of India

Central Public Health and Environmental Engineering Organisation (CPHEEO) (2020) Advisory on landfill reclamation under Swachh Bharat Mission–Urban. Ministry of Housing and Urban Affairs, Government of India

Department of Environment, Forest & Climate Change (DoEFCC) (2023) Legacy waste management Prayagraj, Uttar Pradesh: EPR-based business model. Government of Uttar Pradesh, Prayagraj

Down To Earth (2024) Daily court digest: major environment orders, January 4, 2024. Down To Earth Magazine

Ghosh A, Kartha SA (2024) Landfill biomining of legacy waste dumpsites in India: process, challenges and future perspective. In: Environmental engineering and waste management. Springer, Cham

Ghosh N, Hazra T, Debsarkar A (2022) Biomining: a sustainable solution for reclamation of open landfills in India. Res J Chem Environ 27(1):141–152. https://doi.org/10.25303/2701rjce1410152

Gupta P, Arora S (2021) Policy and regulatory framework for biomining of legacy waste in India. Environ Policy Law 51(2):123–130
Kumar S, Kumar R (2019) Leachate characterization and management in biomining projects: a case study. J Environ Manag 231:337–344
Ministry of Housing and Urban Affairs (MoHUA) (2024) ₹860.35 Cr approved under SBM-U 2.0 to boost urban swachhata infrastructure. Press Information Bureau. https://pib.gov.in/PressReleasePage.aspx?PRID=2029153
Mishra V (2025) Mahakumbh 2024: NGT curious about sudden disappearance of 1.4 million tonnes of Prayagraj's legacy waste. Down To Earth Magazine
Ministry of Environment, Forest and Climate Change (MoEFCC) (2024) Draft Solid Waste Management Rules, 2024. Government of India
Ministry of Environment, Forest and Climate Change (MoEFCC) (2016) Solid Waste Management Rules, 2016. Government of India
Ministry of Environment, Forest and Climate Change (MoEFCC) (2022) Plastic Waste Management (Amendment) Rules, 2022. Government of India
Mohan S, Joseph CP (2020) Biomining: an innovative and practical solution for reclamation of open dumpsites. In: Kalamdhad AS (ed) Recent developments in waste management. Springer, pp 167–178
Mohan V (2024) Centre proposes new rules to manage solid waste across the country with effect from October 1 next year. The Times of India
Ministry of Road Transport and Highways (MoRTH) (2023) Policy guidelines on the use of inert material in the construction of national highways. Government of India
National Green Tribunal (NGT) (2024a) Principal bench order. NGT, New Delhi
National Green Tribunal (NGT) (2024b) Akash Vashishtha v. State of Uttar Pradesh & Others. Principal Bench, National Green Tribunal, New Delhi
National Green Tribunal (NGT) (2023) Lokesh Kumar Khurana v. State of Uttar Pradesh & Others. Principal Bench, National Green Tribunal, New Delhi
Pawade LN, Zanzad RA, Qureshi MN et al (2022) Biomining of legacy waste management. Int J Eng Res Mech Civil Eng (IJERMCE) 9(6):1–7
Press Information Bureau (PIB), Delhi (2022) Lakshya Zero Dumpsite: Government of India approves ₹403.77 Cr legacy waste remediation proposal of Gujarat
Prayagraj Municipal Corporation (PMC) (2024) Reply by R-3 & R-4 Prayagraj Nagar Nigam in OA No. 1181 of 2024: Neeraj Tiwari v. State of Uttar Pradesh
Rai S (2024) NGT ropes in multiple agencies for waste disposal oversight. The Times of India
Rai S (2023) NGT summons Meerut municipal commissioner. The Times of India
Saliba M, Daou A, Kallab A, Mouawad LR (2024) A review of the Miyawaki method: key findings and recommendations. Department of Landscape Design and Ecosystem Management, American University of Beirut. https://www.aub.edu.lb/natureconservation/Documents/miyawaki%20brief.pdf
Shahab S, Anjum M (2022) Solid waste management scenario in India and illegal dump detection using deep learning: an AI approach towards sustainable waste management. Sustainability 14(23):15896. https://doi.org/10.3390/su142315896
Singh R (2022) Toolkit: legacy waste management and dumpsite remediation to support Swachh Bharat Mission 2.0. Ministry of Housing and Urban Affairs, Government of India
Sogarwal G (2025) Legacy waste site remediation in Gorakhpur city: environmental implications and a pathway to green credits. Int J Creat Res Thought 13(1):101–108
Thakkar AR, Semwal N, Gargava P (2023) Biomining of legacy waste dump sites in India—an overview. Int J Ecol Environ Sci 49(2):117–124
Zhou C, Gong Z, Zhang Y (2014) Life-cycle inventory and impact evaluation of mining municipal solid waste landfills. Environ Sci Technol 48(5):2920–2927

Part II
Gravel and Soil-Like Fraction

Chapter 6
Reuse of Gravel and Soil-Like Material Derived from Landfill Mining: Characteristics, Contaminants, and Safeguards

Mohit Somani, Manoj Datta, and G. V. Ramana

6.1 Introduction

Landfill mining (LFM) is increasingly recognized as a viable strategy for addressing the dual challenges of growing urban waste and shrinking land availability. In India and many other developing nations, unengineered legacy dumpsites have expanded rapidly over the decades, leading to environmental degradation and land resource constraints (Somani et al. 2018; Roy et al. 2025). The term "legacy dumpsite" refers to sites where MSW has accumulated for over 10–15 years (Somani 2025). As urbanization continues, these sites not only pose a threat to public health and groundwater quality but also occupy valuable land that could be repurposed for productive use. LFM offers a promising opportunity to reclaim land while simultaneously recovering resources from old waste deposits. Among the various outputs generated during landfill mining, the gravel and soil-like material (GSLM), typically comprising of minus 30 mm or minus 20 mm fraction of municipal solid waste (MSW), constitutes the largest mass of the recovered material. However, the safe reuse of GSLM is restricted by concerns regarding its contamination potential.

The GSLM fraction, usually less than 30 mm in particle size, may visually resemble natural soil but often contains a complex mixture of residual organics, heavy metals, soluble salts, and other pollutants in a matrix of soil and gravel (Parida et al. 2024). These contaminants originate from decades of indiscriminate waste disposal, including domestic, commercial, and industrial refuse. Without proper characterization and treatment, the reuse of GSLM particularly in geotechnical and

M. Somani (✉)
School of Infrastructure, Indian Institute of Technology Bhubaneswar, Khurda, Odisha, India
e-mail: mohitsomani@iitbbs.ac.in

M. Datta · G. V. Ramana
Department of Civil Engineering, Indian Institute of Technology Delhi, New Delhi, India

M. Somani et al. (eds.), *Landfill Mining of Legacy MSW: Global Insights and Indian Experience*, Springer Transactions in Civil and Environmental Engineering, https://doi.org/10.1007/978-981-95-3959-8_6

geoenvironmental applications such as embankments, backfilling, and land reclamation could lead to long-term environmental risks, including leachate generation, groundwater contamination, and bioaccumulation of toxic metals.

The present study focuses on the contamination behavior and reuse challenges associated with GSLM obtained from landfill mining operations in India. Samples of aged MSW and GSLM were collected from three large dumpsites in Delhi, and subjected to a series of laboratory investigations to assess their physical characteristics, contamination levels, and leaching behavior. This comprehensive evaluation includes particle size distribution, organic matter content, total and leachable heavy metal concentrations, and visual leachate analysis. The findings from this study are critically compared against regulatory reuse standards. To mitigate these risks, the study proposes protective measures such as engineered sealing layers, leachate collection systems, and surface capping, especially for applications in shallow embankments and deep excavated pits. Additionally, a range of treatment technologies were reviewed to determine their feasibility for large-scale deployment in resource-constrained settings.

6.2 Materials and Methodology

6.2.1 Sample Collection

Samples of total legacy MSW and GSLM were collected from three dumpsites of Delhi where landfill mining was underway. Total MSW were collected by excavating trial pits of 1 m × 1 m (approx.) up to a depth of 3–4 m. GSLM samples were collected directly from the trommels. The detailed procedure of sample collection is described in Somani et al. (2020) and Parida et al. (2024).

6.2.2 Laboratory Investigations

Compositional analysis and Particle size distribution—The compositional analysis was performed on as-received samples and after washing as per the procedure outlined by Somani et al. (2022). First, samples were washed with tap water to remove fine particles stuck to larger items such as plastic or stones. The washed material was then left to air dry under the sun for 4–5 days. Once dried, the MSW was passed through a 20 mm sieve to separate it into fine (<20 mm) and coarse (>20 mm) fractions. The coarse portion was manually sorted into categories such as plastic, textile, paper, glass, wood, gravel, and miscellaneous items. The quantity of dry MSW used for this analysis ranged from 120 to 140 kg, which meets the ASTM D5231–92 (2016) guideline that suggests using at least 91–136 kg of sample for reliable results.

Particle size distribution was assessed for composite samples collected from 7 to 9 trial pits at each site. About 200 kg of sample per site was subjected to both dry and wet sieving using sieves with openings of 80 mm, 20 mm, and 4.75 mm.

Organic content—The moisture content of the GSLM was first measured after drying it at 60 °C in a thermostatically controlled oven for 48–60 h until constant weight was achieved. The organic content was estimated by heating the dried sample at 550 °C in a muffle furnace for 3–4 h to burn off organic matter.

Total heavy metals: The total heavy metals in GSLM were determined by digesting it with aqua regia (mixture of HNO_3 and HCl in a molar ratio of 1:3) following Somani et al. (2019).

Prior to digestion, GSLM samples were ground to a fine powder (less than 0.075 mm) using a ball mill. A small amount (0.2–0.5 g) of the ground sample was mixed with acid and digested in a microwave system (Multiwave GO 7000, Anton Paar) at 180 °C for 20 min, with a heating rate of 10 °C per minute. The digest was filtered through Whatman No. 42 filter paper and diluted to 50 mL with ultrapure water. The metals—such as Cr, Ni, Cu, Zn, As, Se, Cd, Pb, and Mn—were quantified using inductively coupled plasma mass spectrometry (ICP-MS).

Leaching potential—Single batch leaching test according to the Swedish standard method (SS-EN 12457-2, 2003) was performed. This was adopted on GSLM by the previous investigators Kaartinen et al. (2013). The test involved mixing 10 g of GSLM with 100 mL of deionized water (solid-to-liquid ratio of 1:10) and agitating the mixture at 100 rpm for 24 h using a rotary shaker. The resulting suspension was allowed to settle before being filtered with 2.5 μm Whatman No. 42 filter paper. The filtrate was further centrifuged at 8000 rpm for 10 min, and the final solution was analyzed for heavy metals using ICP-MS. Sulfate and chloride concentrations were measured using the turbidimetric and argentometric titration methods, respectively, as per APHA guidelines (2012).

Release of dark-colored leachate: The intensity of the color of dark-colored leachate (1:10 dilution) from GSLM was measured in Platinum-Cobalt Scale (PCU) using a Lovibond Tintometer.

6.3 Results and Discussions

6.3.1 Characteristics of Excavated Waste

The findings from the compositional analysis of legacy MSW are summarized in Table 6.1. A significant proportion, approximately 67–80% of the total excavated waste consisted of fine materials smaller than 20 mm, such as sand, silt, clay, and fine gravel. The coarser fraction (>20 mm) primarily included construction debris like stones, bricks, and concrete chunks, which accounted for 12–27% of the waste. Combustible materials, including plastic, wood, and textiles, made up a relatively

Table 6.1 Compositional analysis of total MSW samples excavated from legacy dumpsites

Particle size	Constituents	Range (in %)	Values reported in the literature (%)
<20 mm	Finer materials (fine gravel + soil-like material)	67–80	43–75
>20 mm	C&D waste (oversized stones, debris)	12–27	2.0–30
	Combustible materials (plastics, wood, textiles etc.)	2.5–7	14–35
	Others (glass, metals, coconut shell)	0.9–7	1.0–16

Fig. 6.1 Components of total MSW collected from all three dumpsites a, b, and c

small share, ranging from 2.5% to 7.0%. Other components such as glass, metals, and coconut shells were observed in the range of 0.9–7.0%.

When compared with data from previous studies (Krook et al. 2012), the proportion of combustibles found in this study is notably lower. This lower fraction could be attributed to the widespread and informal resource recovery system in India, where waste pickers or "Kabadiwalas" often collect high-value items like plastics and metals before they reach the landfill. Figure 6.1 illustrates the various constituents of the aged MSW from all four sites studied.

The results of particle size distribution analysis are provided in Table 6.2. The data shows that a majority of the excavated waste falls into the soil-like category, specifically materials finer than 4.75 mm. The proportion of soil-like material

Table 6.2 Particle size distribution of total MSW samples excavated from legacy dumpsites

Particle size	Classification	Before washing	After washing
>80 mm	Cobbles + boulders (%)	5.0–15	1.0–8.0
20–80 mm	Coarse gravel (%)	15–22	8.0–15
4.75–20 mm	Fine gravel (%)	18–25	12.0–20
0–4.75 mm	SLM (%)	42–52	59–70

Table 6.3 Elemental composition (expressed in % oxides form) of GSLM determined by XRF

	Size of fraction (mm)			
Parameters (%)[a]	0–30	6–30	0–6	Background soil (Delhi silt)
SiO_2	43–49	46.7	41.0	57–62
CaO	11.5–12	14.0	11.5	5–6
Fe_2O_3	8–9	8–9	5–9	4–5
Al_2O_3	7–8	8	3–8	15–16.5

[a]Adapted from Datta et al. (2021); Parida et al. (2024)

(comprising sand, silt, and clay) ranged between 42% and 52% in samples processed using dry sieving, and between 59% and 70% when wet sieving was applied. These values are relatively higher than those reported in previous studies by Somani 2025, Singh et al. (2017), and Singh and Chandel (2020) for the same sieve threshold.

The higher yield from wet sieving is likely due to the removal of fine particles adhering to coarser fragments during the washing process. This results in a more accurate transfer of fines into the SLM fraction, thereby increasing its percentage in the overall sample.

Table 6.3 presents the results of elemental composition of GSLM (expressed in the form % oxides) performed by XRF. From the XRF results, it can be concluded that bulk of the GSLM consist of SiO_2 (41–49%) which is comparable to the background soils of the study areas (57–62%). GSLM has shown high percentage of CaO in comparison to the background soils which can be attributed to the presence of C&D waste. The elevated CaO content and variable Fe/Al oxides may influence both geotechnical behavior and contaminant retention properties, which are critical considerations for evaluating the reuse potential of GSLM in geoenvironmental applications.

6.3.2 *Reuse of GSLM in Earthfills: Contaminants of Concern*

The following contaminants of concern have been observed from the reuse prespective:

- High organic content.
- Elevated heavy metals concentration.
- High soluble salts.
- Release of dark-colored leachate.

Table 6.4 Organic content of GSLM

Sample description		Organic content (%)
Size of fraction (mm)	0–30	10–12
	6–30	6–12
	0–6	5–18
Background soil		0.8–1
Regulatory limits[a]		1–5

[a]Adopted from Somani et al. (2018)

6.3.2.1 High Organic Content

The organic matter content in the partially air-dried GSLM samples ranged between 5% and 18% across all sites, aligning well with the range reported by other investigators (Kurian et al. 2003; Singh and Chandel 2020; Chandana et al. 2021) for Indian dumpsites. The finest fraction of GSLM consistently showed the highest concentration of organic matter.

Notably, the organic matter in GSLM was found to be 5–15 times higher than that of natural background soils (0.8–1.0%), indicating significant contamination (Table 6.4). Furthermore, the measured organic contents exceed the typical thresholds (1–5%) prescribed by national and international guidelines for the use of soil in earthfill applications (Somani et al. 2018). This excess suggests a high potential for long-term settlement due to continued organic degradation, making untreated GSLM unsuitable for structural fills.

6.3.2.2 Elevated Concentrations of Heavy Metals (Total)

Total concentration of heavy metals represents the concentration of metals in GSLM in all possible forms (water soluble, acid soluble, reducible, residual, etc.).

The heavy metal content was significantly higher in the GSLM than the background soils. High content of Pb, Cr, and Zn can be attributed to the disposal of batteries, fluorescent lamps, chromium sludge and other electronic items. Disposal of paints, insecticides, pharmaceuticals waste may be responsible for the high content of Cu. Cr and Zn were found to be 5–10 times higher in the GSLM than the background soils in most of the cases. Cu was found to be 20–30 times, whereas Cd and Pb were found to be 10–40 higher than the background soils (Table 6.5). Ni, As, and Mn were found to be 2–5 times higher in the GSLM than the background soils.

When compared to the Flemish guideline values for unrestricted reuse, most metals slightly exceed the thresholds. However, all metal concentrations remain

Table 6.5 Concentration of total heavy metals and its comparison with local soil and reuse standards

Sample specifications		Total concentration of heavy metals (mg/kg)							
		As	Cd	Cr	Cu	Mn	Ni	Pb	Zn
Size of fraction (mm)	0–30	1.5–5.3	0.3–2.8	103–312	81–493	225–400	44–81	20–98	148–392
	6–30	1.1–8.2	0.2–2.2	168–274	53–246	185–455	53–107	5–94	99–500
	0–6	4.5–12	0–1.8	185–420	206–599	275–480	68–163	24–110	139–696
Background soil		2.5–3.5	0.05–0.1	17–25	14–20	35–100	20–25	5–6.5	44–67
Unrestricted use (Flemish)		19	2.6	91	135	NA	56	120	529
Restricted use (German)		150	10	600	600	NA	600	1000	1500

Adopted from Somani et al. (2020)

within the German regulatory limits for restricted reuse (e.g., with cover or encapsulation).

6.3.2.3 Leaching of Soluble Salts

The leaching of contaminants from GSLM upon water infiltration can lead to subsoil and groundwater contamination. Therefore, understanding the leaching behavior is crucial for reuse assessment.

The majority of inorganic fraction of total soluble solids consist of sulfates and chlorides. A high concentration of chloride in water is usually considered a major source for the contamination of groundwater. The major sources of sulfates in leachate are decomposition of organic matter, soluble waste, such as ash, detergents, and inert waste, such as sediments of the dredged river. Sulfates in the GSLM were found to be 958–8971 mg/kg, whereas chlorides were found to vary between 1694 and 4098 mg/kg.

The heavy metals in the leachate from GSLM and water extract from background soils are listed in Table 6.6. It can be seen that except cadmium, all the metals were found to be 4–8 times higher in leachate from GSLM than the water extract of background soils. Cadmium was found to exhibit extremely high concentration (57–96 times higher than the background soils).

Compared to reuse standards:

- German regulations (LAGA, 2012): Chloride and sulfate concentrations exceed limits for unrestricted use; in some cases, sulfate concentrations qualify for restricted use (with surface sealing).
- Dutch regulations (VROM, 2007): Chlorides and sulfates generally exceed limits for open reuse but remain within thresholds for isolated reuse.

Table 6.6 Concentration of leachable salts and heavy metals and their comparison with local soil and reuse standards

Sample specifications		Soluble salts (mg/kg)		Total concentration of heavy metals (mg/kg)							
		SO_4^{2-}	Cl^-	As	Cd	Cr	Cu	Mn	Ni	Pb	Zn
Size of fraction (mm)	0–30	2128–6585	1780–3124	0.03–0.23	0.3–2.8	0.03–0.07	0.02–0.88	0.09–1.71	0.01–0.25	ND	0.01–0.52
	6–30	958–6691	1694–2716	0.03–0.15	0.2–2.2	0.03–0.05	0.01–0.89	0.05–2.05	0.01–0.20	0.02	0.35–0.75
	0–6	3277–8971	3099–4098	0.04–0.23	0.06–0.07	1.68–2	0.75–2.50	0.35–2.50	0.34–0.45	0.94–1.25	1.18–3.75
Background soil		300–350	200–250	0.04–0.06	0.001	0.36–0.52	0.25–0.52	ND	0.13–0.21	0.03–0.06	0.22–0.67
No restrictions at all[a]		500	100	0.1	0.02	0.15	0.5	NA	0.4	0.2	1
Sealing of surface[a]		6000	1500	0.5	0.05	1	2	NA	1	1	4
Open reuse[b]		1730	616	0.9	0.04	0.63	0.9	NA	0.44	2.3	4.5
Isolated reuse[b]		20,000	8800	2	0.06	7	1	NA	2.1	8.3	14

[a]German standards—LAGA 2012
[b]Dutch standards (VROM 2007)—soil quality decree 2007

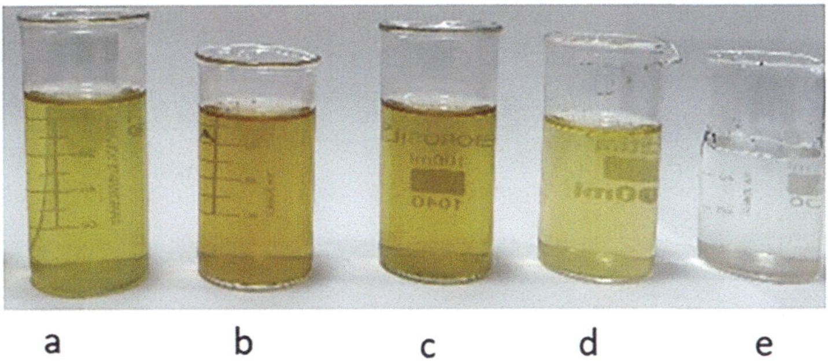

Fig. 6.2 Visual appearance of dark-colored leachate released by GSLM

6.3.2.4 Release of Dark-Colored Leachate

Although limited literature exists on the color of leachate released from GSLM, visual inspections at the dumpsites revealed the formation of dark brown to black leachate pools at the base of the dumps. Such leachate, if generated after reuse of GSLM in low-lying areas or in embankments, can aesthetically pollute surrounding water bodies or reduce public acceptance of groundwater.

- The color intensity of leachate from GSLM samples was found to range from 205–925 PCU, significantly higher than the 25–40 PCU observed in background soils.
- Visual observations (Fig. 6.2a–d) show leachate varying from yellowish to dark brown, whereas background soil leachate (Fig. 6.2e) was clear.

Such coloration may not necessarily indicate chemical hazard but can hinder acceptability and perception of water quality by the public.

6.3.3 Reuse of SLM (<4.75 mm) as Compost

To identify the reuse potential of SLM as compost in agricultural applications, following characteristics of SLM with their associated concerns have been discussed:

- Physicochemical characteristics (pH, electrical conductivity, and total soluble solids)
- Total organic carbon and other nutrients (N, P, and K)
- Total heavy metals

The results of all the parameters listed above have been compared with the international and national regulatory standards of compost.

Table 6.7 Comparison of physicochemical parameters of SLM with compost standards

Sample description		pH	Conductivity (dsm^{-1})	Total soluble solids (mg/kg)
SLM		7.1–8.5	1.2–4.8	8000–25,000
Compost standards	India[a]	6.5–7.5	4	–
	China[b]	–	–	12,000

[a]SWM Rules, 2016
[b]BMBLF, 2012

6.3.3.1 Physicochemical Characteristics

The physicochemical profile of SLM covering pH, electrical conductivity, and total soluble solids was evaluated and compared with reference compost standards (Table 6.7).

The pH of SLM varied between 7.1 and 8.5. In certain samples, this value slightly surpassed the upper compost limit of 7.5 recommended for India. Electrical conductivity values for most samples met the Indian benchmark of 4 dSm^{-1}, although occasional exceedances were recorded. This elevated conductivity aligns with the presence of higher soluble solids, as reflected in the corresponding total soluble solids range. Notably, the soluble solids content in many SLM samples was above the 12,000 mg/kg limit stipulated under Chinese compost regulations (BMBLF, 2012).

6.3.3.2 TOC and Other Nutrients

The concentrations of TOC, total nitrogen, total phosphorous, and total potassium in the SLM, along with their comparison against Indian compost standards, are summarized in Table 6.8.

SLM from all examined dumpsites exhibited TOC levels below the 12% threshold prescribed by Indian compost regulations. This deficiency is likely linked to the mixing of fine inorganic debris originating from construction and demolition waste, street sweepings, drain silt, and miscellaneous refuse into the recovered material Hölzle et al., (2022). In contrast, nutrient parameters such as TN, TP, and TK generally complied with national standards. The findings suggest that, due to its low TOC, SLM cannot be directly utilized as compost in agriculture without carbon enrichment, though it may hold potential for certain non-agricultural uses.

6.3.3.3 Total Heavy Metals

Heavy metal concentrations in SLM were determined following triacid digestion (HNO_3: H_2SO_4: $HClO_4$ = 10: 1: 4) and benchmarked against compost quality standards for various reuse categories (Table 6.9).

The analysis reveals that SLM fails to meet compost quality requirements for organic farming due to elevated concentrations of certain metals. When compared

Table 6.8 Comparison of TOC and nutrients of SLM with compost standards

Sample description	TOC (%)	TN (%)	TP (%)	TK (%)
SLM	3.0–8.7	0.75–1.40	0.65–0.85	0.34–1.70
Compost standards (India)	>12	>0.8	>0.4	>0.4

Table 6.9 Total heavy metals in SLM and compost standards

		Organic farming	Agricultural		Non-agricultural applications
Heavy metal (mg/kg)	SLM	TOA (Austria)[a]	European standards[b]	MSW Rules, 2016 (India)[c]	European standards[b]
Cd	0.5–3.5	0.7	0.7–1.5	5	1.5–5
Cr	65–422	70	50–100	50	100–250
Cu	105–788	70	25–200	300	100–500
Pb	16–338	45	45–150	100	150–1000
As	3.5–7.8	–	15	10	–
Ni	25–70	25	22,920	50	50–100
Zn	170–1000	200	75–500	1000	400–1800
Hg	0.55–4.7	0.4	0.2–1	0.15	44,317

[a]Technical Office of Agriculture, Austria, B4–3040/2001/325284/Mar/A2 Annex 2; June 2004
[b]Brinton, 2000
[c]MSW Rules, 2016

to European guidelines, SLM generally qualifies for non-agricultural applications, except in instances where chromium and copper slightly exceed the upper permissible limits. However, for agricultural applications under European standards, SLM is unsuitable.

Under MSW rules (2016) of India, cadmium, arsenic, and zinc remain within the acceptable range. Conversely, chromium, copper, and lead often surpass the permissible limits, with nickel showing marginal exceedance. These exceedances limit the feasibility of using SLM as compost, particularly for applications involving soil amendment in agriculture.

6.4 Safeguards—Protective Measures

As established in Sect. 6.3, direct reuse of GSLM/SLM is not recommended due to the presence of organic matter, heavy metals, leachable salts, and dark-colored leachate all of which pose potential environmental and health hazards. This section outlines protective measures necessary to mitigate these risks when GSLM/SLM is considered for reuse in bulk earthfilling applications and as a soil conditioner.

6.4.1 Use of GSLM as an Earthfill

Some of the most common applications of GSLM in earthfilling work include:

- In embankments and shallow fills for reducing the consumption of natural soil.
- Reclaiming deep pits (previously degraded by past excavations) by filling them with GSLM, thus restoring the original ground level.

Amongst the above, earthfilling by GSLM in deep pits remain high-risk areas that have the potential to significantly impact the subsurface environment, especially when improper disposal practices are employed.

6.4.1.1 Shallow Fills and Embankments: Provision of Sealing Layers

When GSLM is reused in shallow fills or embankments, especially in areas with high rainfall or shallow groundwater, protective measures become essential to prevent leachate infiltration and groundwater contamination.

Recommended engineered controls include:

- Low-permeability sealing layers (e.g., compacted clay, geosynthetic liners) on the top and side slopes to limit water infiltration and isolate contaminants (Fig. 6.3).
- Leachate drainage layers, composed of coarser materials and embedded perforated pipes, to collect and divert percolating water to surface drains or treatment systems.
- Vegetative cover systems to minimize surface runoff and soil erosion.

Such systems are particularly crucial in hydrogeologically sensitive zones or areas near water bodies. When properly implemented, they:

- Enhance the environmental safety of GSLM reuse.
- Support rapid site remediation and restoration.
- Prevent long-term pollutant migration into underlying soil and aquifers.

6.4.1.2 Deep Fills for Reclamation of Excavated Pits: Engineered Containment

The reuse of GSLM for filling deep excavated pits is high-risk due to the potential proximity of the groundwater table. If contaminants migrate downward, they can directly pollute aquifers and compromise water quality (Fig. 6.4).

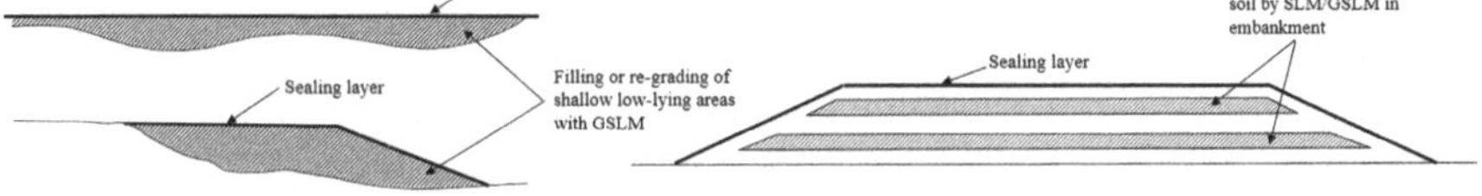

Fig. 6.3 Use of GSLM in low-lying areas with protective measures

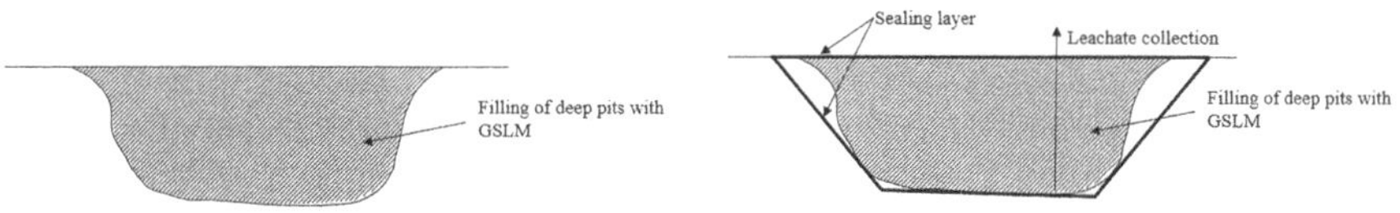

Fig. 6.4 Use of GSLM as an earthfills for deep pits

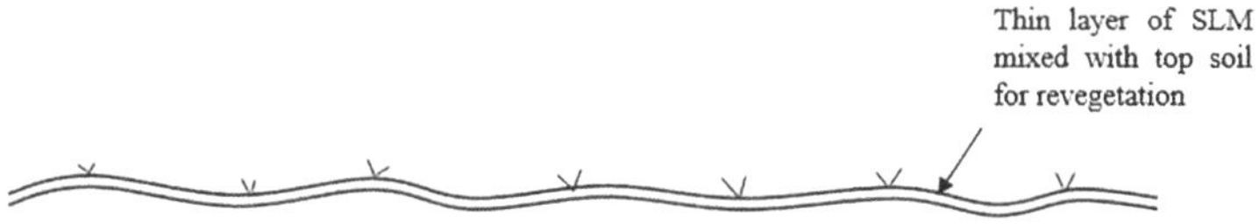

Fig. 6.5 Use of SLM for re-vegetation for non-edible crops

To minimize these risks, robust containment systems are mandatory, typically including:

- Base and side liners made of compacted clay or geomembranes (Fig. 6.5).
- Leachate collection systems with drainage pipes to intercept and remove contaminated liquids.
- Surface sealing layers to limit vertical infiltration from precipitation.

However, it is important to note that such engineered systems can be technically complex and cost-intensive, often rendering the practice economically unviable for large-scale deployment. Therefore, filling deep pits with GSLM without proper containment is strongly discouraged, especially in locations with shallow or vulnerable groundwater regimes.

6.4.2 Use of SLM as Soil Conditioner

6.4.2.1 Surface Applications: Minimal Impact; Root Uptake Concerns

When SLM is applied in thin surface layers (e.g., landscaping, soil conditioning, or re-vegetation), the associated risks are relatively low due to the following reasons:

- The applied thickness is typically limited to 15–30 cm, allowing contaminants to become diluted and blended with native soil (Fig. 6.3).
- Natural attenuation processes such as bioturbation, seasonal wetting and drying, and microbial activity help reduce contaminant concentrations over time.
- Downward leaching of contaminants is minimal, as the unsaturated zone (vadose zone) acts as a natural filter.

However, one notable concern is plant-mediated contaminant uptake. If SLM is used for growing edible crops, heavy metals may be translocated to above-ground plant parts, posing a risk to the food chain. This issue can be addressed by:

- Proper blending SLM with top soil to reduce contaminant concentrations.
- Avoiding the cultivation of edible crops in SLM-amended soils.
- Promoting non-edible or ornamental species for ecological restoration.

6.5 Safeguards—Viability of Treatment Options

For landfill mining to be successful, it is important to use treatment methods that are affordable and suitable for handling large volumes of GSLM. Since the final product is not expected to have high market value, it will have to compete with cheaper natural and recycled materials, using expensive or highly complex technologies is generally not practical particularly in regions where remediation infrastructure is not well-developed.

The treatment also needs to be done quickly and outside the landfill site (ex situ), as GSLM is usually obtained after sorting through excavated waste. These processes might be carried out in specially prepared areas and could resemble some in situ treatments used for porous soils. Fast processing is preferred so that long-term storage or space use is not required. A process is considered "fast" if it can be completed within months or a few years, rather than stretching over decades.

The selected method should also directly address the contamination, particularly by removing harmful substances like heavy metals and salts from the material. This is essential to make the treated material safe for reuse without ongoing monitoring. If the treated material still requires strict control to avoid environmental risk, its market value would likely decrease.

To identify the most suitable options, a basic evaluation was conducted using a simple scoring approach based on how well each method meets the key goals mentioned above (see Table 6.10). First, commonly used technologies for heavy metal and non-metal cleanup, as listed in the land contamination and risk manual (LCRM) Options Appraisal Matrix, were considered. Among them, only flushing or leaching showed some promise for being a low-cost ex situ method with potential to recover usable SLM. However, this method has drawbacks, it can create large volumes of contaminated water that would need additional treatment or safe disposal.

Other globally recognized methods with strong research support were also reviewed Hölzle et al., (2025). Among these, bioremediation (biological stabilization of metals or salts) was found to be less suitable. On the other hand, electrokinetic treatment and phytoremediation showed more promise. Although electrokinetic methods also generate some contaminated liquids, the volume is expected to be much lower than with flushing or leaching.

In summary, treating large quantities of mixed landfill-derived material remains a major challenge. Scaling up such treatments is complex due to the nature of the material and limitations in current technology.

Table 6.10 Screening of remediation options for treating GSLM/SLM[a]

		Low cost/ tech?	Rapid	Source treatment	GSLM/SLM recovery?
Typical technologies listed in LCRM (EA 2020)	Cover systems	Yes	Yes	No	No
	Hydraulic barriers	Yes	Yes	No	No
	In-ground barriers	Yes	Yes	No	No
	Excavation and disposal	Yes	Yes	No	No
	Natural attenuation	Yes	No	Yes	Yes
	Soil flushing / leaching	Yes	Possibly	Yes	Yes
	Chemical oxidation	Possibly	Yes	No	No
	Permeable reactive barriers	No	No	Yes	Possibly
	Soil washing	No	Possibly	Yes	No
	Hydraulic binders	Yes	Yes	No	No
	Vitrification	No	Yes	No	No
	Incineration	Yes	Yes	No	No
	Thermal desorption	No	Yes	Yes	Possibly
Additional technologies Hölzle et al., (2025)	Electrokinetic extraction	Possibly	Possibly	Yes	Yes
	Phytoremediation	Yes	Possibly	Yes	Yes
	Bioremediation (stabilization)	Yes	Possibly	No	No

[a]Adopted from Somani et al. 2023

6.6 Conclusions

This study comprehensively examined the characteristics and environmental risks associated with GSLM derived from landfill mining activities at three legacy dump-sites in Delhi. The findings revealed that although GSLM visually resembles natural soil, it contains significant levels of organic matter, heavy metals, and leachable salts, along with the potential to release dark-colored leachate. These contaminants substantially exceed background soil values and often surpass regulatory thresholds for unrestricted reuse, thereby posing environmental and public health concerns if reused without adequate safeguards.

The high organic content indicates instability and the potential for long-term settlement, while elevated heavy metal concentrations and high concentration of soluble salts highlight risks related to groundwater contamination and bioaccumulation. Additionally, the release of aesthetically unpleasant leachate may hinder public acceptance of reuse initiatives. Therefore, direct reuse of untreated GSLM in construction and land reclamation is not recommended.

On assessing the reuse of SLM as compost, the total organic carbon was found to be lower than the minimum required value; however, by and large the nutrient levels (N, P, and K) satisfy the requirements prescribed in the regulatory thresholds.

Elevated level of total heavy metals in the SLM does not allow its direct usage in agricultural applications.

To enable safer applications, this study proposes engineered protective measures and evaluates treatment technologies for large-scale remediation. Among the reviewed options, electrokinetic treatment and phytoremediation showed potential for reducing contamination levels; however, the feasibility of large-scale applications is yet to be established. Overall, the research underscores the need for stringent assessment and site specific reuse guidelines to ensure environmentally responsible valorization of GSLM in landfill mining projects.

References

ASTM D5231–92 (2016) Standard test method for determination of the composition of unprocessed municipal solid waste

Chandana N, Goli VSNS, Mohammad A, Singh DN (2021) Characterization and utilization of landfill-mined-soil-like-fractions (LFMSF) for sustainable development: a critical appraisal. Waste Biomass Valoriz 12:641–662

Datta, M., Somani, M., Ramana, G. V., & Sreekrishnan, T. R. (2021). Feasibility of re-using soil-like material obtained from mining of old MSW dumps as an earth-fill and as compost. *Process Safety and Environmental Protection*, 147, 477-487.

Hölzle, I., Somani, M., Ramana, G. V., & Datta, M. (2022). Heavy metals in soil-like material from landfills–Resource or contaminants?. *Journal of Cleaner Production*, 369, 133136.

Hölzle I, Somani M, Ramana GV, Fiener P, & Datta M (2025). Sequential washing vs. thermal treatment: heavy metal management in soil-like material derived from landfills. Journal of Environmental Management, 395. https://doi.org/10.1016/j.jenvman.2025.127833

Kaartinen T, Sormunen K, Rintala J (2013) Case study on sampling, processing and characterization of landfilled municipal solid waste in the view of landfill mining. J Clean Prod 55:56–66

Krook J, Svensson N, Eklund M (2012) Landfill mining: a critical review of two decades of research. Waste Manag 32(3):513–520

Kurian J, Esakku S, Palanivelu K, Selvam A (2003) Studies on landfill mining at solid waste dumpsites in India. In: Proceedings Sardinia, vol 3, pp 248–255

Parida D, Ramana GV, Datta M (2024) Investigation on trommeled legacy waste from full-scale mining of old dumpsites: suitable for valorization or scientific disposal? J Environ Manag 356:120580

Roy S, Somani M, Farooq SH (2025) Microbially induced calcite precipitation for improving the geoenvironmental performance of landfilled mined soils. Indian Geotech J:1–14

Singh A, Chandel MK (2020) Effect of ageing on waste characteristics excavated from an Indian dumpsite and its potential valorisation. Process Saf Environ Prot 134:24–35

Singh R, Vivek JM, Vishwakarma N et al (2017) Management of a dump site with mounds of legacy wastes: a proposal for land recovery and material recovery at Pirana dumpsite. In: International conference on integrated solid waste management practices in developing countries, 11–12 April. NEERI, Nagpur

Somani M (2025) Overcoming dumpsites rehabilitation challenges in India: landfill mining versus microbial methane oxidation systems. Indian Geotech J:1–45

Somani M, Datta M, Ramana GV, Sreekrishnan TR (2018) Investigations on fine fraction of aged municipal solid waste recovered through landfill mining: case study of three dumpsites from India. Waste Manag Res 36(8):744–755

Somani M, Datta M, Ramana GV, Sreekrishnan TR (2019) Leachate characteristics of aged soil-like material from MSW dumps: sustainability of landfill mining. J Hazard Toxic Radioact Waste 23(4):04019014
Somani M, Datta M, Ramana GV, Sreekrishnan TR (2020) Contaminants in soil-like material recovered by landfill mining from five old dumps in India. Process Saf Environ Prot 137:82–92
Somani M, Datta M, Ramana GV, Hölzle I, Sundaram R, Sreekrishnan TR (2022) Effect of depth of landfill on the characteristics of soil-like material of aged waste: a case study of Bhalswa dumpsite, India. J Mater Cycle Waste Manag 24(5):1902–1912
Somani M, Harbottle M, Datta M, Ramana GV, Sreekrishnan TR (2023) Identification and assessment of appropriate remediation management techniques for the recovery of soil-like material produced in landfill mining. J Environ Manag 348:119300

Chapter 7
Composition and Characterization of Legacy MSW Obtained from Old Dumpsites

Kasinathan Muthukkumaran and M. Ambal Devi

7.1 Introduction

The rapid increase in global solid waste generation can be attributed to population growth, urbanization, and economic development. By 2025, global waste production is anticipated to rise to 2.3 billion tons annually, with India alone contributing approximately 56 million tons per year (CPCB 2019; Tyagi et al. 2018). Much of this waste is managed through conventional landfill practices, where around 35 million tons are dumped annually, resulting in massive landfills and severe environmental issues. These challenges include greenhouse gas emissions, unpleasant odors, insect infestations, leachate generation, and groundwater contamination Devahi et al. (2022). The environmental and social challenges posed by landfills emphasize the significance of implementing sustainable waste management practices.

In response to these challenges, landfill mining (LFM) has emerged as a promising approach to modern waste management. LFM not only mitigates environmental pollution but also offers benefits such as land reclamation, resource recovery, landfill capacity expansion, and cost savings in landfill closure (Hernández Parrodi et al. 2018). By recovering valuable materials and generating fuel resources, LFM adheres to the principles of a circular economy and supports the remediation of dumpsites. In the last three decades, more than 60 LFM projects have been initiated worldwide (Zhou et al. 2015), demonstrating the transformative potential of this technique.

One of the key outcomes of LFM is the recovery of SLM from MSW, also known as fine particles, which constitute 38–78% of the extracted material. These fractions

K. Muthukkumaran (✉) · M. Ambal Devi
Department of Civil Engineering, National Institute of Technology Tiruchirappalli, Tiruchirappalli, Tamil Nadu, India
e-mail: kmk@nitt.edu

M. Somani et al. (eds.), *Landfill Mining of Legacy MSW: Global Insights and Indian Experience*, Springer Transactions in Civil and Environmental Engineering, https://doi.org/10.1007/978-981-95-3959-8_7

exhibit potential for reuse in diverse applications, including construction, road sub-base layers, and agricultural soil amendments. Numerous studies have explored the reuse potential of MSW derived from LFM, highlighting its applicability as earthfill material for on-site and off-site projects due to favorable geotechnical properties (Hernández Parrodi et al. 2018; Masi et al. 2014). Nevertheless, the presence of toxic (heavy metals) elements considerably limits its application, especially in sensitive applications such as road construction and embankments. Techniques such as washing and blending have been proposed to reduce heavy metal content and enhance usability (Hanson et al. 2013; Wanka et al. 2017), though large-scale reuse remains underexplored.

MSW has also been evaluated for its potential as compost material, given its nutrient content, which is sometimes adequate for agricultural applications (Alvarenga et al. 2015). However, studies have indicated variability in nutrient levels, often requiring nutrient supplementation to meet compost standards (Zhou et al. 2015; Prechthai et al. 2008). The existence of heavy metals further complicates its use in agriculture, as these can affect soil quality and plant growth (Mukhopadhyay et al. 2016).

This study seeks to assess the composition and characteristics of MSW retrieved from the Ariyamangalam dumpsite in Trichy, India, to evaluate its potential for reutilization. The research concentrates on analyzing the physicochemical properties of water extracts from MSW and local soil to determine compliance with drinking water standards. It further examines contamination indices to categorize pollution levels and investigates geotechnical characteristics to assess suitability for earthfill applications. Finally, the research evaluates the applicability of MSW as earthfill and compost material, ensuring compliance with regulatory standards to promote sustainable waste management practices.

7.2 Materials and Methodology

7.2.1 Site Overview

The study focused on excavated MSW retrieved from the Ariyamangalam landfill, situated on the periphery of Tiruchirappalli city, India. This dumpsite, in operation since 1965, covers 47.5 acres and accumulates approximately 100 tonnes of MSW daily, 210 tonnes monthly, and 520 million tonnes annually. The biomining process at the site generates various by-products, including SLM, compost, and construction debris, with only SLM selected for this research. Five batches of MSW were collected, dried under natural sunlight, and sieved through a 2 mm filter to remove contaminants like broken glass, paper, and textiles, in accordance with IS 2720-4: 1985. For comparative analysis, local soil was gathered from an area more than 500 meters from the dumpsite at a depth of 20 cm to ensure minimal anthropogenic influence on the sample. Figure 7.1 illustrates: (a) The retrieval process of MSW

Fig. 7.1 (**a**) Retrieval of MSW (**b**) Sun dried MSW

from the Ariyamangalam dumpsite, and (b) The sun-drying of the retrieved MSW, preparing it for subsequent analysis.

7.2.1.1 Morphological and Elemental Study

The surface morphology of MSW was examined using a scanning electron microscope (SEM). SEM images were captured at varying magnifications under a 3 kV electron beam acceleration potential. Additionally, energy-dispersive X-ray spectroscopy (EDS) was performed on the MSW samples to analyze their elemental composition.

7.2.1.2 Mineralogical Study

X-ray diffraction (XRD) analysis was conducted on the MSW samples to determine their mineralogical composition. The test was performed using a Cu-Kα radiation source at a scanning range of 2θ from 10° to 80°.

7.2.2 *Physicochemical Parameters*

The pH of SLM was determined by diluting 75 ml of distilled water into 30 g of the sample, sieved through a 425 μ sieve, as per IS 2720: Part-XXVI. Total dissolved solids were quantified by mixing 10 g of soil, passing through a 2 mm IS sieve, with 100 ml of distilled water. After 15 h of shaking, the clear filtrate was passed through Whatman filter paper, and 50 ml of the filtrate was dried at 110 °C in an oven. The residue was weighed and expressed as a percentage, in accordance with IS 2720: Part-XXI. Organic content was evaluated by heating a 50 g oven-dried sample in a muffle furnace at 440 ± 40 °C for 1 h to achieve complete ashing. The remaining ash was used to calculate the organic content, following ASTM: D2974. Total soluble sulphates were analyzed as per IS 2720: Part-XXVII. Additionally, the color intensity of water extracts was measured on the Platinum-Cobalt Scale (PCU) using a 1:10 dilution ratio.

7.2.3 Heavy Metals Concentration and Leachability

Heavy metal concentrations were assessed through two digestion methods. Aqua regia (1:3 nitric acids to hydrochloric acid) was used to evaluate suitability as earth-fill material, while a tri-acid mixture (10:4:1 Nitric acid, Perchloric acid, and Sulfuric acid) analyzed its potential for compost applications. Metals analyzed included various heavy metals using inductively coupled plasma mass spectrometry (ICP-MS). Indices including the Contamination factor, geo-accumulation index, degree of contamination, and enrichment factor were computed for categorization. Leachate and toxicity characteristics were determined using the batch leaching method in accordance with Swedish Standard EN 12457-2: 2003. This method, known for its simplicity and reliability, was adopted to evaluate the environmental hazards posed by MSW.

7.2.4 Geotechnical Properties

The geotechnical properties of MSW and native soil were assessed through various standard tests. Particle size distribution was analyzed using wet sieving and hydrometer methods, adhering to IS 2720 (Part—4): 1985. Specific gravity was determined as per IS 2720 (Part—3): 1980. The Free Swell Index test was conducted following IS 2720: Part 40. Optimum Moisture Content (OMC) and Maximum Dry Density (MDD) were calculated from the peak of the smooth curve obtained through the light compaction test, in accordance with IS 2720: Part VII. Direct shear tests were carried out in line with IS 2720 (Part—13): 1986, using specimens measuring 25 mm in height and 60 × 60 mm in size. The testing protocol involved soaking samples for 24 h and applying normal stresses of 50, 100, and 150 kPa, with a constant shear displacement rate of 0.30 mm/min. These comprehensive analyses provided insights into the geotechnical behavior of the materials under varying conditions.

7.2.5 Nutrient Levels

The total organic carbon content was calculated using loss-on-ignition results and a conversion factor (Iglesias Jiménez and Pérez García 1992). Total nitrogen, phosphorus, and potassium concentrations were assessed in line with the Fertilizer (Compost) Order of India (FCO 1985).

7.2.6 *Quality Management and Oversight*

All experiments were conducted in triplicate to ensure accuracy, with results showing variations exceeding 15% discarded and re-evaluated. High-purity water served as the control to minimize deviations, ensuring consistent quality control throughout the analyses.

7.3 Results and Discussions

7.3.1 *Makeup and Particle Distribution of MSW*

The waste product of LFM predominantly consists of SLM, which has a dark hue and makes up the largest proportion of the recovered material. At the Ariyamangalam land fill, 78% of the SLM was found to be finer than 2 mm during wet sieving, results align with earlier studies conducted at the dumpsite in Kadapa and Delhi (Somani et al. 2019, 2020). In contrast, dry sieving yielded 72% of SLM. The makeup of the MSW from the Ariyamangalam site includes glass, plastics, paper, fabrics, and textiles, with a significant fraction (55–80%) being combustible materials.

7.3.1.1 Morphological and Elemental Study

The SEM analysis of MSW obtained from the dumpsite reveals a heterogeneous microstructure characterized by irregularly shaped particles of varying sizes. The images show a combination of angular, flaky, and amorphous structures, suggesting a mixture of inorganic and organic components. Larger angular fragments indicate the presence of mineral residues or construction debris, while finer granular particles around these fragments suggest mixing with organic matter or secondary mineral formation. The rough surface texture of the particles points to prolonged degradation and physical interactions in the dumpsite environment. Aggregated clusters observed in certain regions may result from microbial activity or chemical bonding, with porous areas potentially facilitating water retention and microbial colonization. Bright spots visible in the images could represent metallic or high-density materials, possibly remnants of electronic waste or industrial sources, highlighting the potential for heavy metal contamination. Figure 7.2 shows the SEM images of MSW.

The EDAX spectrum of MSW obtained from dumpsites provides valuable insights into its elemental composition. Figure 7.3 illustrates the EDAX results of MSW. The spectrum reveals the presence of key elements such as Silicon (Si), oxygen (O), Aluminum (Al), Calcium (Ca), Iron (Fe), and Potassium (K), which are indicative of mineralogical components like silicates and oxides. The prominent peaks of Si and O suggest the dominance of silicate minerals, commonly found in soil and construction debris.

Fig. 7.2 SEM images of MSW

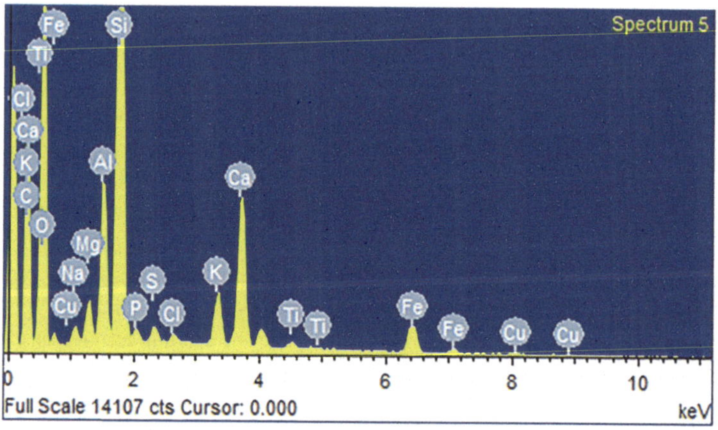

Fig. 7.3 EDAX result of MSW

Secondary elements such as Sodium (Na), Magnesium (Mg), Sulfur (S), Phosphorus (P), Chlorine (Cl), and Titanium (Ti) were also detected, which could arise from mixed waste, including organic degradation and industrial inputs. The detection of heavy metals like Copper (Cu) and Iron (Fe) points to potential contamination sources, possibly from electronic waste or industrial residues. The presence of calcium could be attributed to construction materials like cement or lime-based compounds.

This elemental composition reflects the heterogeneous nature of the MSW, which is typical for dumpsite material comprising diverse waste sources. The high proportion of silicates and oxides highlights its potential mineralogical stability, while the presence of heavy metals underscores the need for further chemical and environmental analysis to evaluate potential risks and remediation strategies. Such an elemental profile is crucial for understanding the material's suitability for reuse, its environmental impact, and its role in soil reinforcement or other applications.

7.3.1.2 Mineralogical Study

The XRD analysis of the MSW obtained from dumpsite provides insight into its crystalline structure and mineralogical composition. The XRD pattern, as shown in Fig. 7.4, exhibits several distinct peaks in the 2θ range corresponding to d-spacing values between approximately 1 and 10 Å. These peaks indicate the presence of crystalline phases within the material.

The prominent peaks at lower d-spacing values, particularly around 3–4 Å, suggest the dominance of silicate minerals such as quartz and feldspar, which are commonly found in natural soils and construction debris. The presence of broad, less-defined peaks may indicate amorphous phases or poorly crystalline materials, likely resulting from organic matter degradation or waste decomposition. Peaks at higher d-spacing values may correspond to clay minerals such as kaolinite or montmorillonite, which are often found in soil and could result from the mixing of natural soil with waste material.

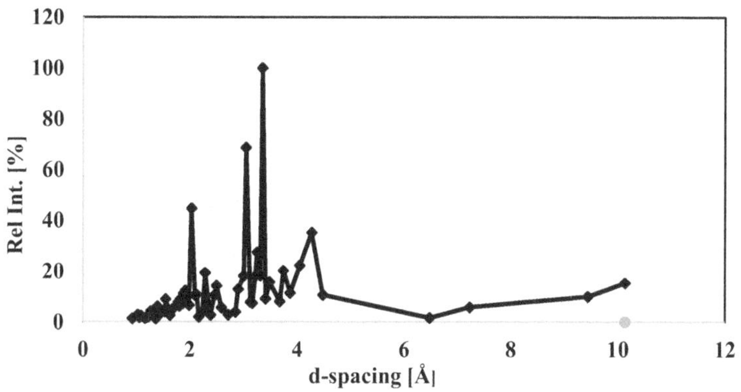

Fig. 7.4 XRD Result of MSW

7.3.2 Physicochemical Parameters

The detailed physicochemical characteristics of the MSW and local soil are presented in Table 7.1 and Table 7.2. Both materials fall within the humic phase, with pH values of 8.06 for MSW and 7.32 for local soil. The humic phase, characterized by pH values ranging from 7 to 8.5, indicates full breakdown of readily degradable organic matter and slower degradation of remaining material, reducing toxic release (Bozkurt et al. 1999). Chemical oxygen demand (COD) values were found to be 133.4 mg/L for MSW and 25.1 mg/L for local soil, further supporting the classification of MSW as part of the humic phase.

The discharge of dark-hued leachate from waste disposal sites is a frequent occurrence, often resulting in its accumulation in low-lying areas as leachate ponds. This leachate is known to be toxic, contributing to the contamination of both underground water and surface bodies of water. Additionally, the suitability of water for consumption diminishes. As color intensity increases, making the management of such leachate critical. Investigating its characteristics is essential for exploring reusable options for MSW. The leachate from MSW showed a greater color depth of 380 PCU compared to 25 PCU for local soil, reflecting MSW's higher organic content. The color is mainly due to humic and fulvic acids. This indicates that the local soil is less organic in nature, a finding consistent with trends observed in previous investigations. Studies from various dumpsites in Indian cities have reported color intensities ranging from 343 to 819 PCU, due to the existence of organic material within the waste. Nonetheless, the color strength of leachate can be diminished through methods like dilution, air exposure, and heating at approximately 550 °C.

Total soluble solids were measured at 1%, reflecting a relatively low level of dissolved organic and inorganic substances in the waste. In contrast, the ash content of the waste was notably high, at 91%, signifying that most of the waste consists of inorganic substances. This high ash content can limit the biodegradation potential of the waste, reducing the likelihood of methane generation but may pose challenges related to landfill stability and leachate formation. The total soluble sulfates present in the waste were 2%, which could contribute to the production of Hydrogen sulfide and lead to the acidification of the leachate, further complicating waste management efforts.

Table 7.1 Physicochemical characteristics of MSW

Parameters	Result
COD	133.4 mg/L
Color	380 PCU
In situ water content	22%
pH	8.06
Total soluble solids	1%
Organic content	19%
Ash content	91%
Total soluble sulphates	2%

Table 7.2 Comparison of heavy metals and leachable concentrations from MSW across various dumpsites

		EC	Color	OC	TSS	Cl^-	SO_4^{2-}	Cr	Ni	Cu	Zn	As	Fe	Cd	Pb
Leachable parameter	pH	dsm^{-1}	PCU	%	mg/kg	mg/L	mg/L	µg/L	µg/L	µg/L	µg/L	µg/L	µg/L	µg/L	µg/L
Present study															
Ariyamangalam landfill, Trichy	8.06	1.47	380	19.6	9360	184.7	369.7	162.6	89.2	41.48	562.3	1	6.55	1.2	10.82
Local soil, Trichy	7.32	0.11	25	1.5	562	49	38.7	43.33	27.6	3.89	43.47	BDL	1.16	BDL	4.51
Research on national waste disposal sites															
Okhla dumpsite, Delhi (Somani et al. 2019)	7.25	2.29	368	6.5	16,000	440	529	194.5	120	138	180	23.5	–	6.95	30.3
Ukkayyapalli landfill, Kadapa (Somani et al. 2019)	7.9	2.45	410	18.3	900	529	652.5	182	11,231	238	292	79	–	6.5	28
Jawaharagr landfill, Hyderabad (Somani et al. 2019)	7.75	0.6	819	11.4	20,250	240	292	188.5	114.5	118	143	58	–	6.95	32.5
Existing studies on international dumpsites															
Finland (Kaartinen et al. 2013)	–	–	–	–	–	850	7100	150	100	146	850	–	–	100	150
Estonia (Burlakovs et al. 2016)	–	–	–	–	–	–	–	88.5	36.5	276.5	1650	5.2	41,750	1.17	302.5
Bangladesh (Karim 2014)	–	–	–	–	–	–	–	45.6	5.37	25.71	91.61.	ND	–	–	–
Sweden (Jani et al. 2016)	–	–	–	–	–	2185	–	54	33.7	684	488	1.7	8.11	0.6	65.3
Standards for regulatory criteria concerning reclamation															
Germany standards of technical bulletin (standard: 6.11.2003)				1	–	10	50	15	40	50	100	10	–	2	20
				2	–	20	150	30	50	50	100	10	–	2	40
				3	–	40	300	75	100	150	300	40	–	5	100
				4	–	150	600	100	100	200	400	50	–	5	100

7.3.2.1 Heavy Metals Concentration and Leachability

The reuse potential of MSW from the Ariyamangalam dumpsite was evaluated through a detailed assessment of its heavy metal concentrations. Table 7.2 presents the heavy metal and leachable concentrations in MSW from different dumpsites. The metal levels in MSW were found to be significantly higher than those in local soil, with Lead (10.82 μg/L), Chromium (162.6 μg/L), Nickel (89.2 μg/L), Copper (41.48 μg/L), and Zinc (562.3 μg/L), along with a very low concentration of cadmium. According to the standards outlined in Table 7.2, the leachable concentrations of Copper, Arsenic, Cadmium, and Lead were deemed suitable for unrestricted usage. However, Nickel and Sulfates (369.7 mg/L) were acceptable only under sealed surface conditions, while the concentrations of Chromium, Zinc, and Chlorides (184.7 mg/L) were classified as unsafe, even with sealed surfaces. Considering its reuse potential, the MSW was found suitable for construction applications as a soil or waste material, with or without a liner. However, its application should be limited in sensitive land areas due to the elevated levels of specific metals such as Chromium, Nickel, and Zinc (Gupta et al. 2021; Somani et al. 2019).

7.3.3 *Organic Content and Moisture Content*

Organic and moisture content are interconnected parameters that provide insights into the degradability of waste. In Indian dumpsites, the average moisture content typically ranges from 8% to 25%, with the current study observing a moisture content of 22%, consistent with previous research (Gupta et al. 2021; Joseph et al. 2003; Song et al. 2003). The in situ water content of 22% reflects a moderate moisture level that can influence the decomposition of organic matter. This moisture level also affects the compaction and stability of the waste and may contribute to leachate formation, which can lead to contamination of nearby environments. The organic content was found to be 19%, indicating a predominance of non-organic materials such as plastics, metals, and glass, which are more resistant to biodegradation. Compared to local soil, MSW has 3 to 4 times higher organic content, with the local soil containing 6.5% organic matter. According to Indian standards, the permissible organic content for soil used as subgrade material is between 1% and 3%, whereas in the United States, the limit is higher, set at 5% (Somani et al. 2018).

7.3.4 *Geotechnical Properties*

Table 7.3 presents the geotechnical characteristics of MSW. MSW used in current study consists of sand, silt, and clay in proportions of 79%, 19%, and 2%, respectively. Figure 7.5 depicts the particle size distribution curves for MSW. This indicates that the waste composition predominantly comprises coarser particles, such as

Table 7.3 Geotechnical Properties of MSW

Parameters	MSW
D_{10}	0.04
D_{30}	0.34
D_{60}	1.5
Cu	37.5
Cc	1.92
Specific gravity	2.11
Soil classification	SM
Free swell index	20%
Optimum moisture content (%)	26
Maximum dry density (g/cc)	1.57
Cohesion (kg/cm^3)	Negligible
The angle of internal friction φ	41.32°

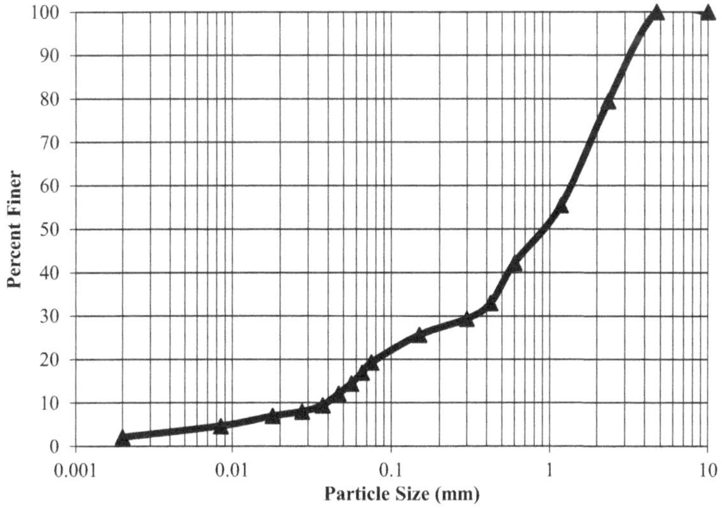

Fig. 7.5 Particle size distribution of MSW

those from construction and demolition waste. Previous investigations of national dumpsites reported MSW grain size distributions of 58–75% sand, 20–36% silt, and 4–6% clay (Somani et al. 2018), consistent with this study except for the minimal clay content. Studies of international dumpsites found sand and silt proportions ranging from 69% to 80% and 12% to 21%, respectively, aligning with the current findings. The low clay content in the retrieved MSW reflects higher degradation rates and the landfill's age (Zhong et al. 2018; Giardina et al. 2001). The specific gravity of MSW in this study was 2.11, comparable to the Jawahar Nagar landfill in Hyderabad, but lower than the 2.5–2.7 range reported for national dumpsites, likely due to higher organic content. A low specific gravity indicates a high presence of organic content.

The plasticity index of MSW was found to be non-plastic, consistent with findings from national and international landfills (Quaghebeur et al. 2013). In contrast, local soil exhibited a plasticity index of 10 and a liquid limit of 28%, categorizing it as low-plasticity soil.

Figure 7.6 illustrates the compaction curve of MSW. Experimental results showed the maximum dry density and optimum moisture content of MSW to be 1.57 t/m^3 and 26%, respectively, compared to 1.87 t/m^3 and 11% for local soil. The lower dry density of MSW can be attributed to its heterogeneous composition, fibrous materials, and degradation age. National dumpsites showed higher maximum dry densities (1.34–1.68 t/m^3), while the Ariyamangalam results align with the Sydney landfill (1.13–1.24 t/m^3) (Fatahi et al. 2015).

Figure 7.7 presents the graph of the Direct Shear Test. A direct shear test was conducted on compacted MSW finer than 2 mm, revealing dilatant behavior during volume changes. The angle of shear resistance measured was 44°, consistent with

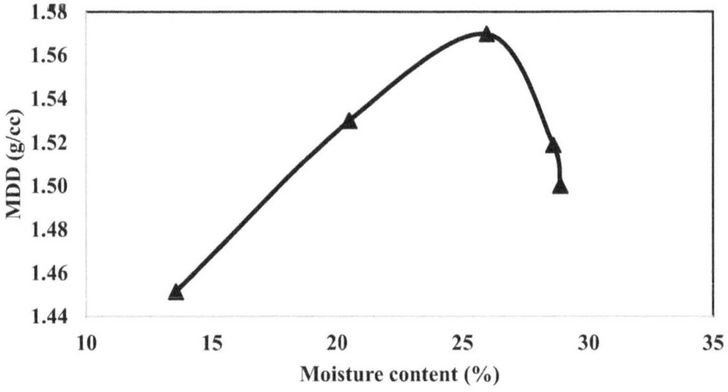

Fig. 7.6 Compaction curve of MSW

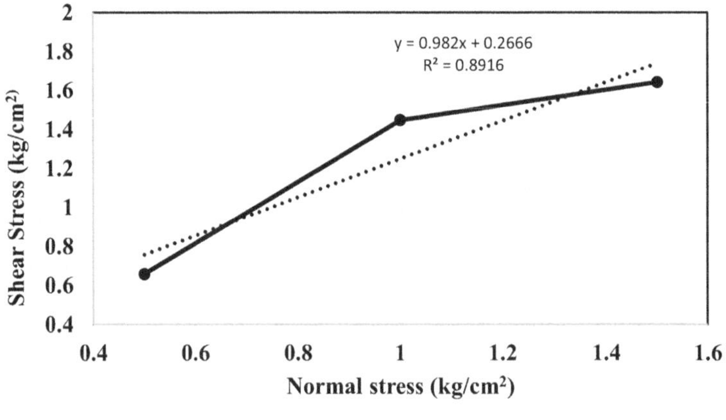

Fig. 7.7 Direct shear test graph

studies at Michigan's monofill, USA (Zekkos et al. 2013). Local soil exhibited a shear resistance angle of 18.26°, while national dumpsites showed values between 34 and 38°, and international dumpsites ranged from 12° to 42° (Zekkos et al. 2013; Song et al. 2003). The impact of organic content on shear resistance was minimal (Somani et al. 2018).

The shear strength characteristics of MSW from the Ariyamangalam dumpsite indicate its potential for reuse as on-site landfill applications. However, its use as fill material for off-site applications is unsatisfactory without adequate measures to prevent groundwater contamination. Applications in areas with minimal long-term settlement risks, such as parking lots or golf courses, are possible with proper treatment, including water washing, incineration, and additive blending (Oettle et al. 2010; Wanka et al. 2017). Phytoremediation techniques could also remediate contamination, particularly heavy metals. Care should be taken in regions with high groundwater tables.

7.3.5 Concentration of TOC

Experimental results indicate that the MSW contains 1.09% TN, 0.55% TP, and 0.82% TK, meeting the minimum requirements for compost material (SWM Rules 2016). However, the TOC concentration in the MSW was found to be lower than the standard's prescribed levels, likely due to the high presence of inorganic materials such as drain silt and construction debris (Saha et al. 2010). Despite the reduced TOC, its concentration is consistent with values reported for the Hyderabad dumpsite (Somani et al. 2018) and higher than those recorded at other dumpsites, including Delhi, Finland, and Beijing (Rong et al. 2017; Mönkäre et al. 2016; Somani et al. 2018). Regarding TK concentration, the MSW from Ariyamangalam exhibited higher levels compared to the Hyderabad and Yingchun landfills, but lower concentrations than those reported for the Delhi, Nanqijia Town, and Nonthaburi dumpsites (Rong et al. 2017).

Except for TOC, the concentrations of TN, TP, and TK in the MSW provide a sufficient nutrient base for agricultural applications as compost. Due to its lower TOC, the MSW may be more suitable for reuse as compost in non-agricultural applications.

7.3.6 Eco-conscious Reuse Capability

The current study highlights that SLM generally constitutes 45–78% of the MSW recovered from LFM projects. At the Ariyamangalam dump yard, the SLM finer than 2 mm accounts for 72% of the entire portion of MSW recovered. This abundant material presents significant potential for bulk use in infrastructure development projects, such as highway embankments, trench filling, mine reclamation, and filling open pits. Reusing SLM in such applications can help reduce the need for finite and non-renewable resources like river sand. Given its nutrient-rich composition, MSW also shows potential as organic matter for farming uses. However, the presence of heavy metals, leachable contaminants, and dark-colored leachate, which can cause groundwater pollution, raises concerns about its reuse. These challenges require further investigation and innovative solutions from engineers and researchers to enable safe and sustainable reuse.

This study emphasizes the importance of implementing additional safety measures and tailored design strategies to address these risks and enhance the reuse potential of MSW in various applications.

7.4 Conclusion

- The findings from the leaching test show that the water extract from MSW surpasses the permissible limits for drinking water standards concerning total dissolved solids (TDS) and sulfates. Nevertheless, the other physicochemical characteristics of the water extract from MSW adhere to drinking water standards.
- The concentration of total fixed soluble solids indicates that MSW is suitable for use in confined earthfill and construction applications.
- Concerning organic content, MSW surpasses the regulatory limits established by Indian code provisions for earthfill material, with an organic content of 19%. The allowed organic content for subgrade applications is 1%–3% according to Indian standards and 5% under U.S. standards. Therefore, MSW cannot be reused without undergoing treatment to lower its organic content.
- MSW could be appropriate for large-scale reuse in regions with shallower depths and lighter load-bearing demands, like rural roads. However, it is advisable to use liner material beneath the MSW to safeguard nearby surface and subsurface water sources.
- A comparative analysis of MSW confirmed their appropriateness for earthworks. However, the reuse of MSW is restricted in sensitive land applications due to the concentrations of nickel (0.11%) and zinc (0.12%), which exceed the limits set by Swedish guidelines.
- The nutrient content in MSW—1.1% total nitrogen (TN), 0.5% total phosphorus (TP), and 0.8% total potassium (TK)—indicates its potential for composting. However, the absorption of heavy metals by plants must be closely monitored. Additionally, the total organic carbon (TOC) concentration in MSW was found to be below 0.4%, which may necessitate the addition of organic material. As a result, MSW is advised for use in non-edible plants and non-agricultural applications. These results highlight the potential of MSW for controlled reuse in particular applications, while underscoring the importance of appropriate treatment and safety measures to address environmental risks.

References

Alvarenga P, Mourinha C, Farto M, Santos T, Palma P, Sengo J, Morais M-C, Cunha-Queda C (2015) Sewage sludge, compost and other representative organic wastes as agricultural soil amendments: benefits versus limiting factors. Waste Manag 40:44–52. https://doi.org/10.1016/j.wasman.2015.01.027

Bozkurt S, Moreno L, Neretnieks I (1999) Long-term fate of organics in waste deposits and its effect on metal release. Sci Total Environ 228(2–3):135–152. https://doi.org/10.1016/S0048-9697(99)00047-9

Burlakovs J, Kaczala F, Vincevica-Gaile Z, Rudovica V, Orupõld K, Stapkevica M, Bhatnagar A, Kriipsalu M, Hogland M, Klavins M, Hogland W (2016) Mobility of metals and valorization of sorted fine fraction of waste after landfill excavation. Waste Biomass Valoriz 7(3):593–602. https://doi.org/10.1007/s12649-016-9478-4

CPCB (2019) CPCB notice—dumpsite ban

Fatahi B, Khabbaz H, Fatahi B (2015) Chapter 8—improving geotechnical properties of closed landfills for redevelopment using chemical stabilization techniques: a case study on samples of a landfill site in southwest of Sydney. In: Indraratna B, Chu J, Rujikiatkamjorn C (eds) Ground improvement case histories. Butterworth-Heinemann, pp 239–266. https://doi.org/10.1016/B978-0-08-100191-2.00008-3

FCO (1985) No.11-3/83-STU Government of India Ministry of Agriculture and Rural Development (Department of Agriculture and Cooperation)

Giardina C, Ryan M, Hubbard R, Binkley D (2001) Tree species and soil textural controls on carbon and nitrogen mineralization rates. Soil Sci Soc Am J 65:1272–1279. https://doi.org/10.2136/sssaj2001.6541272x

Gupta G, Datta M, Ramana GV, Alappat BJ (2021) MSW incineration bottom ash (MIBA) as a substitute to conventional materials in geotechnical applications: a characterization study from India and comparison with literature. Constr Build Mater 308:124925. https://doi.org/10.1016/J.CONBUILDMAT.2021.124925

Hanson J, Yesiller N, Onnen M, Liu W-L, Oettle N, Marinos J (2013) Development of numerical model for predicting heat generation and temperatures in MSW landfills. Waste Manag 33:1993–2000. https://doi.org/10.1016/j.wasman.2013.04.003

Hernández Parrodi JC, Vollprecht D, Pomberger R (2018) Characterization of fine fractions from landfill mining: a review of previous investigations. Detritus 2:46–62. https://doi.org/10.31025/2611-4135/2018.13663

Iglesias Jiménez E, Pérez García V (1992) Relationships between organic carbon and total organic matter in municipal solid wastes and city refuse composts. Bioresour Technol 41(3):265–272. https://doi.org/10.1016/0960-8524(92)90012-M

Jani Y, Kaczala F, Marchand C, Hogland M, Kriipsalu M, Hogland W, Kihl A (2016) Characterisation of excavated fine fraction and waste composition from a Swedish landfill. Waste Manag Res 34(12):1292–1299. https://doi.org/10.1177/0734242X16670000

Joseph K, Esakku S, Palanivelu K, Selvam A (2003) Studies on landfill mining at solid waste dumpsites in India. In: Proceedings Sardinia'03, ninth international landfill symposium

Kaartinen T, Sormunen K, Rintala J (2013) Case study on sampling, processing and characterization of landfilled municipal solid waste in the view of landfill mining. J Clean Prod 55:56–66. https://doi.org/10.1016/J.JCLEPRO.2013.02.036

Karim R (2014) Assessment of heavy metal contamination from municipal solid waste open dumping sites in Bangladesh. J Hydrol Environ Res (JHER) 2:41–49. https://doi.org/10.1504/IJEWM.2017.10005182

Mönkäre TJ, Palmroth MRT, Rintala JA (2016) Characterization of fine fraction mined from two Finnish landfills. Waste Manag 47:34–39. https://doi.org/10.1016/J.WASMAN.2015.02.034

Mukhopadhyay S, Masto RE, Cerdà A, Ram LC (2016) Rhizosphere soil indicators for carbon sequestration in a reclaimed coal mine spoil. CATENA 141:100–108. https://doi.org/10.1016/J.CATENA.2016.02.023

Oettle, N., Matasovic, N., Kavazanjian, E., Rad, N., & Conkle, C. (2010). Characterization and placement of municipal solid waste as engineered fill

Prechthai T, Padmasri M, Visvanathan C (2008) Quality assessment of mined MSW from an open dumpsite for recycling potential. Resour Conserv Recycl 53(1–2):70–78. https://doi.org/10.1016/J.RESCONREC.2008.09.002

Quaghebeur M, Ben L, Geysen D, Nielsen P, Pontikes Y, Van Gerven T, Spooren J (2013) Characterization of landfilled materials: screening of the enhanced landfill mining potential. J Clean Prod 55:72–83. https://doi.org/10.1016/j.jclepro.2012.06.012

Devahi P, Rathod D, Muthukkumaran K (2022) Building material synthesization using municipal solid waste incineration ash: A state-of-the-art review. Environ Technol Rev 11:33

Rong L, Zhang C, Jin D, Dai Z (2017) Assessment of the potential utilization of municipal solid waste from a closed irregular landfill. J Clean Prod 142:413–419. https://doi.org/10.1016/j.jclepro.2015.10.050

Saha JK, Panwar N, Singh MV (2010) An assessment of municipal solid waste compost quality produced in different cities of India in the perspective of developing quality control indices. Waste Manag 30(2):192–201. https://doi.org/10.1016/j.wasman.2009.09.041

Somani M, Datta M, Gupta S, Sreekrishnan TR, Ramana GV (2019) Comprehensive assessment of the leachate quality and its pollution potential from six municipal waste dumpsites of India. Bioresour Technol Rep 6. https://doi.org/10.1016/j.biteb.2019.03.003

Somani M, Datta M, Ramana GV, Sreekrishnan T (2020) Contaminants in soil-like material recovered by landfill mining from five old dumps in India. Process Saf Environ Prot 137. https://doi.org/10.1016/j.psep.2020.02.010

Somani M, Datta M, Ramana GV, Sreekrishnan TR (2018) Investigations on fine fraction of aged municipal solid waste recovered through landfill mining: case study of three dumpsites from India. Waste Manag Res 36:0734242X1878239. https://doi.org/10.1177/0734242X18782393

Song Y-S, Yun J-M, Hong W-P, Kim T-H (2003) Investigation of solid waste soil as road construction material. Environ Geol 44:203–209. https://doi.org/10.1007/s00254-002-0746-1

SWM Rules. (2016). Ministry of environment, forest and climate change notification

Tyagi V, Fdez-Güelfo LA, Alvarez C, Romero Garcia L, Ng J (2018) Anaerobic co-digestion of organic fraction of municipal solid waste (OFMSW): progress and challenges. Renew Sust Energ Rev 93. https://doi.org/10.1016/j.rser.2018.05.051

Wanka S, Münnich K, Fricke K (2017) Landfill mining—wet mechanical treatment of fine MSW with a wet jigger. Waste Manag 59:316–323. https://doi.org/10.1016/J.WASMAN.2016.10.050

Zekkos D, Kabalan M, Syal SM, Hambright M, Sahadewa A (2013) Geotechnical characterization of a municipal solid waste incineration ash from a Michigan monofill. Waste Manag 33(6):1442–1450. https://doi.org/10.1016/j.wasman.2013.02.009

Zhong Z, Chen Z, Yadong X, Yang G, Han X, Ren G (2018) Relationship between soil organic carbon stocks and clay content under different climatic conditions in Central China. Forests 9:598. https://doi.org/10.3390/f9100598

Zhou C, Xu W, Gong Z, Fang W, Cao a (2015) Characteristics and fertilizer effects of soil-like materials from landfill mining. Clean (Weinh) 43:n/a–n/a. https://doi.org/10.1002/clen.201400510

Chapter 8
Contaminants of Concern in Landfill Mining Residues

Supriya Mohanty and Parul Rawat

8.1 Introduction

Globally, more than two billion tonnes of municipal solid waste (MSW) are produced annually. If compressed into standard shipping containers and arranged in a continuous line, they could encircle the Earth 25 times or stretch farther than the round trip to the moon (UNEP 2024). For decades, much of this waste has ended up in landfills, considered a low-cost solution for disposal. However, this practice comes with significant challenges. Valuable resources are wasted, organic waste decomposes to release methane—a potent greenhouse gas—and hazardous substances can leach into the environment, causing local pollution. Additionally, space constraints have become a pressing issue, especially in densely populated areas where landfills often hinder urban development. Despite advancements in waste management and recycling, landfilling remains a dominant practice, leaving behind vast amounts of accumulated waste in old and operational sites, some of which contain materials of significant value (Krook et al. 2012). Landfill mining (LFM) is a practical approach to managing pollution and recovering valuable resources. Over time, waste in landfills can stabilize, making it feasible for extraction and processing. In the case of waste dumps, LFM is often undertaken to remediate sites and prevent ongoing contamination of soil and air. While not a widely adopted practice, its implementation is growing. LFM focuses on excavating and processing old waste from active or closed landfills to reclaim materials, reduce dependency on raw resources, and address environmental concerns effectively (Jain et al. 2023). In recent years, Enhanced Landfill Mining (ELFM) has gained attention, emphasizing resource recovery through advanced technologies like Waste-to-Material (WtM)

S. Mohanty (✉) · P. Rawat
Department of Civil Engineering, Indian Institute of Technology (BHU), Varanasi, India
e-mail: supriya.civ@iitbhu.ac.in

M. Somani et al. (eds.), *Landfill Mining of Legacy MSW: Global Insights and Indian Experience*, Springer Transactions in Civil and Environmental Engineering, https://doi.org/10.1007/978-981-95-3959-8_8

and Waste-to-Energy (WtE) processes. Pilot projects have primarily focused on recovering scrap materials and producing refuse-derived fuel (RDF), while composting organic waste has been explored less frequently. Besides resource recovery, ELFM provides additional benefits such as site remediation and space recovery in landfills. A significant portion of landfill waste is soil-like material (SLM), which often accounts for 50–60% of aged waste. SLM, also known as landfill fines or inert waste, has variable particle sizes and is used in various applications, including landfill cover, noise barrier earth berms, road construction, and backfilling quarries. These practices further contribute to sustainable waste management and the reuse of valuable materials (Hölzle et al. 2022). SLM is unsuitable for compost or fuel but can be used as soil cover, a soil amendment, biocover, or conditioner. Challenges like contaminant leaching require binding in impervious matrices. Combining SLM with other landfill waste at the source could reduce costs and supports sustainability (Goli and Singh 2023). SLM has minimal economic value and high transportation expenses. In certain countries, like India, a market is emerging for compliant SLM, but treating it is costly, with limited success in removing contaminants ((Hölzle et al. 2022).

The primary challenge in reusing landfill-mined residues lies in identifying and understanding the contaminants they contain, such as heavy metals, organic pollutants, and microplastics. Assessing the leachability, mobility, and potential environmental and health risks associated with these contaminants is critical. Addressing these challenges requires the development of sustainable strategies for managing these residues, including mitigation, neutralization, and reuse. Such efforts not only contribute to pollution control and resource recovery but also align with the principles of a circular economy. Effectively tackling these issues is essential for minimizing the hazards posed by landfill mining residues, protecting the environment, and promoting sustainable waste management solutions.

8.2 Types of Contaminants in Landfill Mining Residues

A meta-analysis of heavy metal concentrations in soil-like material (SLM) by (Hölzle et al. 2022) globally included data from 59 landfills, sourced from 23 publications and 17 environmental agency reports. The data covered 59% of studies from Europe, 30% from Asia, and 11% from the USA. Most studies (75%) were recent, with 56% conducted in the last 5 years. Chromium was analyzed in 52 landfills, while mercury was the least (25 landfills). The study found that zinc had the highest median concentration (455 mg/kg) in landfill residues, followed by lead, copper, and chromium. Mercury, cadmium, and arsenic had the lowest medians. Zinc, cadmium, and copper showed frequent extreme outliers, while mercury and lead had fewer. Some samples had concentrations much higher than the median, but these were rare and localized. It was found that metals like chromium and nickel were lower in SLM, while arsenic, lead, cadmium, copper, mercury, and zinc were higher, indicating contamination risks for reuse. Compared to European compost standards, most metals exceeded the limits, especially lead. Zinc, copper, chromium, and cadmium also often exceeded C&D (construction and demolition debris) reuse thresholds.

In addition, this investigation centers on soil-like material (SLM, <4.75 mm) obtained from aged municipal solid waste (MSW) at five historic dumpsites across Delhi, Hyderabad, and Kadapa in India. The analysis revealed distinct levels of heavy metal contamination within the SLM matrix: arsenic and selenium were deemed uncontaminated; nickel and manganese exhibited minimal pollution; chromium showed moderate contamination; zinc and cadmium ranged from moderate to severe pollution levels; and copper and lead were identified as highly contaminated. Elevated concentrations of total soluble solids (5140–25,220 mg/kg) suggest that incorporating SLM into the environment could significantly raise soil salinity and pose a threat to nearby groundwater resources. Furthermore, the organic matter content in SLM, ranging from 4.8% to 24.5%, far exceeds regulatory thresholds for earthworks, contrasting sharply with the lower organic content of 0.9–1.38% found in local soils (Somani et al. 2020). Table 8.1 provides the data of five Indian landfills and international standards to compare with.

In terms of metal recovery, copper concentrations in SLM were closest to ore grades, while others like nickel, lead, and mercury were much lower. Some landfills showed high concentrations of multiple metals, often linked to the disposal of materials like batteries and plastics, suggesting potential for resource recovery in certain cases. Heavy metals in SLM from landfills are compared to natural levels and regulatory limits for reuse in construction and agriculture. However, these limits vary by region and depend on local conditions, with countries like the USA, Canada, France, and Germany setting different thresholds. Zinc, lead, and cadmium often exceed limits, while arsenic, mercury, and copper are less of a concern. While some studies show compliance with compost standards, high metal levels, particularly lead, chromium, nickel, and zinc, often limit SLM's reuse in agriculture (Hölzle et al. 2022).

Studies reveal that fine fractions in landfill mining (LFM) generally contain low metal recovery potential, except in cases where industrial waste (IW) or coarser particle limits (<40 mm) were included. Concentrations of metals like Fe, Al, Cu, Zn, and rare earth elements (REEs) vary by site and particle size. Metals such as Fe, Mg, Zn, and Cu are present in significant amounts (e.g., Fe >10,000 mg/kg) in some studies. Heavy metals tend to accumulate in fine fractions due to their high surface interaction, and their recovery requires reduction to elemental forms. Critical and rare metals (e.g., Co, Nd, Sb, and Li) were found in small concentrations but offer future recovery potential. Improved sorting and processing technologies are vital for optimizing metal yields. Metal recovery from fines, alongside other valuable materials, could make LFM economically viable, potentially reducing mining costs by up to 40% as raw material prices increase. However, leaching rates and site-specific conditions significantly influence metal mobility and recovery feasibility (Parrodi et al. 2018).

While most of the studies have mainly focused on the physicochemical properties of SLM, the thermal analysis has been largely overlooked. Understanding the thermal stability of SLM is crucial for assessing its suitability in various

Table 8.1 Total heavy metal concentration (mg/kg)

			Chromium	Nickel	Copper	Zinc	Arsenic	Selenium	Cadmium	Lead	Manganese
	Okhla		89–230	39–53	140–501	153–326	5.7–7.8	3.2–5.4	0.28–1.2	27–333	295–328
	Bhalswa		147.6–199.6	26.21–49.15	203.39–297.93	406.49–571.2	2.58–4.39	3.42–6.03	1.81–3.76	137.05–161.28	198–232.5
Indian dumpsite	Noida		203–339	59–97	189–372	371–579	4.97–6.72	4.30–5.70	1.15–2.58	66–97	355–408
	Hyderabad		125.85–143.16	52.33–60.38	473.32–574.17	423.54–696.3	4.05–5.71	2.59–3.15	2.55–3.53	134.34–196.42	125–152
	Kadapa		94.85–102.91	51.42–54.65	327.45–343.04	414.06–443.5	7.28–7.67	2.33–2.35	1.98–2.02	141.22–155.31	224–268
	Australia	Compost	400		200	250	20		3	200	
	Belgium		70		90	300			1.5	120	
	Canada (CCME)		210	62	400	700	13		3	150	
	China		300				30		3	100	
International standards	Germany		100		100	400			1.5	150	
	India (BIS)		50	50	300	1000	10		5	100	
	Flemish regulations	General	91	56	135	529	19		2.6	120	
		Construction material (soil)	880	250	375	1250	250		10	1250	
		Fertilizer and compost	250	50	375	900	150		6	300	
		Construction material (waste)	1250	250	375	1250	250		10	1250	

Considered from Chandana et al. (2021); Somani et al. (2020)

applications. Thermogravimetric analysis (TGA) and bomb calorimetry (BC) can help estimate the calorific value (CV) and thermal degradation stability. The presence of organic matter (OM) significantly influences SLM's properties, with higher OM content making it more prone to environmental degradation. Studies have shown that SLM with high OM (6–12%) is vulnerable to disintegration under stress, limiting its use in structural applications and waste-to-energy (WtE) processes due to low CV. Furthermore, the characteristics of SLM, such as size fraction and landfill age, influence its properties, but these factors alone are insufficient for accurate characterization. Unlike inorganic soils, SLM contains organic fibers with a high aspect ratio, making conventional testing methods unreliable. A comprehensive understanding of SLM's OM content and its effects on its overall characteristics is crucial for evaluating its suitability for different applications and promoting sustainable reuse (Goli et al. 2022). Moisture and organic matter are key factors in landfill material processing, influenced by waste type, climate, and operations. Fine fractions retain more moisture (16–54%) and organic content (5–21%), with levels decreasing in older sites due to prolonged decomposition. Drying fines improves processing, recyclability, and energy recovery. Stable organic matter slows degradation, while cover materials like compost or clay impact breakdown rates. These parameters guide processing and reuse decisions (Parrodi et al. 2018).

Landfills, receiving ~42% of the 359 million tonnes of global plastic waste generated annually, act as significant sources of microplastics (MPs) due to the fragmentation of macroplastics under varying conditions such as high temperatures (60–90 °C), fluctuating pH (4.5–9), compaction, physical stress, and limited microbial activity. These MPs, often originating from industrial, household, and wastewater treatment plant sources, act as vectors for hazardous chemicals and pathogens, exacerbating environmental and health risks. These MPs are not only embedded within the fine fractions of waste but also contribute to the composition of landfill-generated leachates (Silva et al. 2021). Research on microplastics in landfill leachates has revealed varying levels of contamination across different regions. Kilponen (2016) first reported microplastics in leachate from a closed landfill in Finland, with concentrations up to 1.1 items/L, though larger particles (>1 mm) may have led to underestimation. The Nordic Council of Ministers (van Praagh et al. 2018) found microplastic levels up to 4.5 items/L in Nordic countries, with common plastics like PE, PA, PVC, PET, and tire rubber, and noted treatment processes reduced these levels. Su et al. (2019) observed higher levels (4.0–13.0 items/L) in Shanghai, China, with variations in microplastic patterns by landfill age. He et al. (2019) reported similar levels (0.4–24.6 items/L), highlighting that landfills are significant sources of microplastics. Xu et al. (2020) found much higher concentrations (291.0 ± 91.0 items/L), using refined methods to detect smaller microplastics (20–100 μm). These studies suggest that landfills, particularly in terms of refuse and leachates, can be substantial contributors to microplastic pollution, with variability depending on factors like landfill age, treatment methods, and sampling techniques, calling for standardized methodologies and long-term monitoring to assess their environmental impact.

8.3 Leachability and Mobility of Contaminants

Leachate composition varies with landfill age, climate, and waste type, necessitating detailed analysis for effective management. It contains organic compounds, heavy metals, pathogens, and salts, with properties like pH, turbidity, and conductivity indicating its condition. Contaminant levels, including BOD, COD, and heavy metals, fluctuate over time, requiring continuous monitoring to determine toxicity and guide treatment methods (Nalladiyil and Babu 2024). To understand its utilization potential, metal leachability should be assessed under different pH conditions, incorporating parameters like interaction time, liquid-to-solid ratio, particle size, and shaking intensity. Additionally, evaluating contaminant migration through SLM is crucial for its application as an environmental remediation material. This can be quantified using the distribution coefficient (k_e), determined through batch, column, and lysimeter tests. Factors affecting k_e include pH, ionic strength, organic matter content, temperature, and particle size. For comprehensive evaluation, nutrient retention, acid neutralizing capacity, and pathogen content should also be examined to assess SLM's toxicity and suitability as a biofertilizer or acid-buffering barrier material (Chandana et al. 2021).

Batch leaching tests on SLM from various Indian dumpsites revealed that heavy metals such as chromium, copper, nickel, zinc, cadmium, and lead exhibit significant leachable concentrations, often exceeding regulatory thresholds. This highlights their potential to contaminate groundwater and ecosystems if untreated. Total soluble solids (TSS) in SLM were found to be exceptionally high, ranging from 4880 to 25,220 mg/kg, compared to local soil standards of 500–1000 mg/kg, indicating a pronounced risk of salinity-driven pollution. The presence of organic content and soluble salts further enhances contaminant mobility, compounded by the release of dark-colored leachate (205–925 PCU), which poses aesthetic and acceptability challenges for water resources. These findings emphasize the need for rigorous pre-treatment measures, such as stabilization, encapsulation, or isolation, to minimize leaching risks before considering SLM for reuse in earthworks or other applications (Somani et al. 2020).

8.4 Health and Environmental Implications

The Ariyamangalam dump yard in Tamil Nadu, India, spanning 47.04 hectares with 1.05 million metric tons of legacy waste, poses significant health and environmental risks. Soil-like material (SLM) recovered from the site showed elevated levels of lead (Pb), chromium (Cr), and zinc (Zn), with pollution indices indicating severe contamination, particularly for Pb. Health risk assessments revealed minimal non-carcinogenic risks but significant carcinogenic risks from hexavalent chromium and nickel, especially for children, with ingestion as the primary exposure route. To mitigate these risks, the study recommends remediation measures such as washing and thermal treatment before SLM reuse. It underscores the need for regulatory frameworks to ensure sustainable landfill management and safe material reuse (Gurusamy and Thangam 2023). Similarly, the Docking Common Landfill in Norfolk, UK, spanning 4.6 hectares and containing municipal solid waste (MSW) from the 1960s–1980s, highlights the widespread health and environmental challenges of landfill mining. Analysis of soil-like material (SLM) from this site revealed elevated levels of heavy metals, including lead (Pb), copper (Cu), zinc (Zn), cadmium (Cd), and chromium (Cr), with concentrations far exceeding UK Soil Guideline Values. Pollution indices identified severe contamination, particularly from Pb and Cu, while fine fractions, which constituted over 50% of the waste, were enriched with metals due to their high surface area. Health risk assessments showed that Pb posed significant non-carcinogenic risks, while Cr and arsenic (As) exceeded acceptable carcinogenic risk thresholds, particularly for residential and agricultural uses. To address these risks, the study emphasizes the need for stringent safety protocols, including air dispersion modeling and controlled handling during landfill mining operations, to minimize environmental contamination and safeguard public health (Zari et al. 2022).

Excavation of landfills significantly increases the release of harmful pathogens into the environment, posing serious health risks. Pathogens like Salmonella, Staphylococcus aureus, and E. coli are found in landfill refuse, surrounding soil, vegetation, and air. Culturable bacteria concentrations can reach up to 1010 CFU/g (CFU: colony forming units), with plant leaves serving as secondary contamination sources. Excavated areas release the highest levels of bioaerosols, with concentrations peaking at 6.56×104 CFU/m^3 downwind. Inhalation of these airborne pathogens is the primary health risk, with exposure reaching hazardous levels up to 500 meters away. This highlights the need for protective measures to control bioaerosol release and reduce health risks for nearby residents. Proper management of excavation and landfill remediation is crucial to minimize microbial contamination (Wu et al. 2024). The Environmental Protection Agency (EPA) offers several pathogen sterilization methods for biosolids, including high temperature (pasteurization, sterilization, incineration), chemical agents (alkalines, phenols, alcohols), and radiation (e.g., gamma rays). However, these methods can also destroy beneficial microorganisms essential for plant growth, such as nitrogen-fixing bacteria (NFB) and phosphorous-solubilizing bacteria (PSB), reducing the effectiveness of landfill mining soil fractions as biofertilizer. Therefore, alternative sterilization methods like composting, anaerobic digestion, lime stabilization, and air drying should be explored to ensure SLM remains safe and effective for biofertilizer use (Chandana et al. 2021).

8.5 Treatments and Mitigation

Municipal Solid Waste (MSW) fines excavated from landfill sites often require advanced treatment before they can be safely and effectively used as geomaterials. These fines typically contain high organic content, hazardous compounds, and heavy metals, which limit their direct application. Organic matter, derived from biodegradable waste and daily cover materials, increases biological activity and decomposition potential, leading to challenges in long-term stability and usability. To address these issues, several treatment methods have been developed, each with specific advantages and limitations. Washing techniques are among the simplest and most cost-effective methods, used to remove contaminants such as organic matter, salts, and heavy metals (Sorlini et al. 2011). Soil washing or wet mechanical separation can significantly enhance the quality of MSW fines, particularly in separating fine mineral particles from other waste streams (Wanka et al. 2017). Blending, another straightforward approach, involves mixing MSW fines with locally available soils to reduce organic content and soluble salts while improving geotechnical properties, making the material suitable for use in engineered fills with moderate settlement tolerances (Oettle et al. 2010). Thermal treatment is an effective way to burn off organic content and chemically stabilize the material, reducing contaminant leaching (Rawat and Mohanty 2021; Somani et al. 2018). However, this method is energy-intensive and requires controlled thermal plants due to the low heating value of MSW fines. Stabilization techniques, such as incorporating binders like cement, lime, or asphalt, immobilize heavy metals and improve the mechanical properties of the material. This approach has been successfully applied to similar waste materials, such as incinerated bottom ash, to achieve solidification and reduce environmental risks (Chen et al. 2008; Cioffi et al. 2011). Emerging bio-based methods offer innovative solutions for treating MSW fines (Prozorov 2015). Biocementation, using microorganisms to induce calcium carbonate precipitation, strengthens the material and has shown promise in soil stabilization and self-healing concrete applications. Similarly, bio-stabilization employs aerobic and anaerobic processes to reduce organic content and stabilize waste. These processes can be enhanced through pre-treatments like particle size reduction, leachate recirculation, and the optimization of moisture, pH, and temperature (Hrad et al. 2013). Studies have demonstrated the feasibility of these methods in reducing organic matter content, stabilizing fines, and evaluating geotechnical properties such as hydraulic conductivity and shear strength (Petrovic 2016). While these techniques show promise, large-scale application faces challenges, including high costs, bulk handling, and the need for controlled environments. Transporting and processing MSW fines, alongside implementing infrastructure for treatment, can make the process economically unfeasible despite its environmental benefits. However, reusing MSW fines is crucial for sustainable waste management, reducing landfill volumes, and addressing environmental concerns. With appropriate infrastructure and funding, these treatment methods could contribute significantly to resource recovery and sustainable engineering practices.

8.6 Limitations of Contamination Studies Related to Landfill Mining (LFM) in India

In India, studies on contamination linked to landfill mining (LFM) face several limitations that reduce their usefulness in planning and decision-making. One of the main problems is the lack of proper documentation about old dump sites—details such as the types of waste deposited, the depth of the landfill, and operational practices are often missing. This makes it difficult to accurately assess the extent and nature of pollution. Most studies mainly focus on surface-level waste, with limited attention to deeper layers, leachate movement, or groundwater contamination. There are also no fixed guidelines for how to take and analyze samples, leading to varied and sometimes unreliable results. Due to limited funding and technical skills, many research efforts do not use advanced tools like geophysical surveys, remote sensing, or laboratory-based contaminant testing. Moreover, most studies only look at basic physical and chemical properties of the waste, without considering the biological, ecological, and health risks that are equally important. Lastly, weak enforcement of environmental regulations and poor coordination between local authorities, pollution boards, and researchers delay action and make data-sharing difficult. Improving contamination studies in India will require better planning, stronger regulations, modern tools, public engagement, and more collaboration between different sectors.

8.7 Future Research Directions

Various landfill mining initiatives worldwide, highlighting their successful applications and outcomes. Studies from China, India, and other regions show how landfill mining can recover valuable resources, promote sustainability, and support circular economy principles. For instance, Qin et al. (China) (Qin et al. 2023) demonstrated that soil-like materials from landfills could be solidified with biomass fly ash binders, enhancing their mechanical properties for backfilling. Goli et al. found that incorporating up to 5% landfill-mined soil-like fractions into polymer composites significantly improved their tensile and flexural strengths. Somani et al. (2022) analyzed landfill waste at Delhi's Bhalswa dumpsite and found higher concentrations of heavy metals with depth, while the North Delhi Municipal Corporation processed 661,454 tonnes of waste into refuse-derived fuel (RDF). Cheela et al. (2021) study showed that RDF from Indian landfills had a calorific value between 10.4 and 21.8 MJ/kg, indicating its potential for renewable energy. Goel and Kalamdhad (2017) demonstrated that degraded MSW could partially replace soil in fired brick manufacturing, with a 20% mix showing the highest durability. The Kumbakonam dumpsite in India successfully reclaimed land through biocapping and RDF processing, establishing a biogas unit and processing unit. Indore Municipal Corporation reclaimed 1.3 million tonnes of waste, utilizing recycled materials for construction and repurposing land for a golf course. These studies illustrate the untapped potential of landfill mining for resource recovery, energy generation, and sustainable development, providing valuable lessons for future waste management strategies. To further enhance the effectiveness and sustainability of landfill mining (LFM), future research should focus on several emerging areas. Advanced contaminant management through novel binding agents and stabilization techniques is essential to mitigate contamination risks. Resource recovery optimization, including advanced sorting, recycling, and metal recovery, especially for rare and critical metals, can improve the economic viability of LFM. Additionally, enhancing waste-to-energy technologies by studying the thermal characteristics of landfill-mined residues will help improve energy extraction. Biological remediation methods, such as bioremediation and phytoremediation, offer eco-friendly solutions for managing contaminants, while health and environmental risk assessments are crucial to understanding the long-term impacts of landfill mining residues. Developing robust policy and regulatory frameworks for LFM, ensuring safe and sustainable practices, and understanding the lifecycle impacts of landfill mining operations, including carbon emissions and energy use, will also be vital. Addressing these areas can make landfill mining more efficient, safer, and sustainable, ultimately contributing to global waste management and resource recovery efforts (Nalladiyil and Babu 2024).

8.8 Conclusions

This chapter has explored the challenges and opportunities associated with contaminants in landfill mining residues (LFMR), particularly focusing on heavy metals, organic pollutants, and other environmental hazards. Soil-like material (SLM) recovered from landfills presents significant reuse challenges due to contamination risks, leachability, and environmental implications. Comprehensive studies highlight the persistent issue of elevated heavy metal concentrations, with elements like lead, chromium, zinc, and cadmium frequently surpassing safe reuse thresholds. Despite these challenges, emerging remediation technologies, advanced sorting techniques, and innovative applications such as polymer composites and waste-to-energy processes present promising pathways for sustainable resource recovery. Health and environmental concerns, including bioaerosol releases and pathogen contamination, underscore the necessity of stringent safety protocols and effective pre-treatment methods. Future advancements in landfill mining (LFM) must focus on optimizing contaminant stabilization, improving resource extraction efficiency, and integrating biological and thermal remediation methods. Establishing robust policy frameworks and regulatory guidelines will also be essential to ensure safe and sustainable operations. By addressing these challenges and leveraging technological innovations, landfill mining can evolve into a pivotal component of global waste management strategies, contributing to circular economy goals, resource conservation, and sustainable development.

References

Chandana N, Goli VSNS, Mohammad A et al (2021) Characterization and utilization of Landfill-Mined-Soil-Like-Fractions (LFMSF) for sustainable development: a critical appraisal. Waste Biomass Valoriz 12(2) Springer:641–662

Cheela VRS, John M, Dubey B (2021) Quantitative determination of energy potential of refuse derived fuel from the waste recovered from Indian landfill. Sustain Environ Res 31:1–9
Chen J-S, Chu P-Y, Chang J-E et al (2008) Engineering and environmental characterization of municipal solid waste bottom ash as an aggregate substitute utilized for asphalt concrete. J Mater Civ Eng 20(6) American Society of Civil Engineers (ASCE):432–439
Cioffi R, Colangelo F, Montagnaro F et al (2011) Manufacture of artificial aggregate using MSWI bottom ash. Waste Manag 31(2):281–288
Goel G, Kalamdhad A (2017) Degraded municipal solid waste as partial substitute for manufacturing fired bricks. Constr Build Mater 155:259–266
Goli VSNS, Singh DN (2023) Valorization of landfill mined plastic waste and soil-like fractions in polymer composites–a comprehensive solution for sustainable landfill mining. J Clean Prod 420:138349
Goli V, Singh P, Singh D et al (2022) Investigations on characteristics of landfill-mined-soil-like-fractions and their dependency on organic matter. Process Saf Environ Prot 162:795–812
Gurusamy S, Thangam R (2023) Potential health risk assessment of contaminants in soil-like material recovered from landfill mining. Environ Monit Assess 195(2):330
He P, Chen L, Shao L et al (2019) Municipal solid waste (MSW) landfill: a source of microplastics?-evidence of microplastics in landfill leachate. Water Res 159. Elsevier:38–45
Hölzle I, Somani M, Ramana GV et al (2022) Heavy metals in soil-like material from landfills–resource or contaminants? J Clean Prod 369:1–8
Hrad M, Gamperling O, Huber-Humer M (2013) Comparison between lab-and full-scale applications of in situ aeration of an old landfill and assessment of long-term emission development after completion. Waste Manag 33(10):2061–2073
Jain M, Ashwani K, Amit K (2023) Landfill mining: a review on material recovery and its utilization challenges. Process Saf Environ Prot 169:948–958
Kilponen J (2016) Microplastics and harmful substances in urban runoffs and landfill leachates: possible emission sources to marine environment. Lathi University of Applied Sciences
Krook J, Svensson N and Eklund M (2012) Landfill mining: a critical review of two decades of research. Waste Manag 32(3). Pergamon: 513–520
Nalladiyil A and Babu GL (2024) Dumpsite remediation through landfill mining and rehabilitation: a circular economy perspective. J Indian Inst Sci Springer: 1–30
Oettle N, Matasovic N, Kavazanjian E et al (2010) Characterization and placement of municipal solid waste as engineered fill. In: Global Waste Management Symposium, San Antonio, TX, 2010
Parrodi J, Höllen D, Pomberger R (2018) Characterization of fine fractions from landfill mining: a review of previous investigations. Detritus 2:46–62
Petrovic I (2016) Mini-review of the geotechnical parameters of municipal solid waste: mechanical and biological pre-treated versus raw untreated waste. Waste Manag Res 34(9):840–850
Prozorov T (2015) Magnetic microbes: Bacterial magnetite biomineralization. Semin Cell Develop Biol 46:36–43
Qin Z, Jin J, Liu L et al (2023) Reuse of soil-like material solidified by a biomass fly ash-based binder as engineering backfill material and its performance evaluation. J Clean Prod 402. Elsevier:136824
Rawat P, Mohanty S (2021) Experimental investigation on MSW fine mixed with fibers: fiber reinforced waste. J Hazard Toxic Radioact Waste 25(3) American Society of Civil Engineers:04021009
Silva ALP, Prata JC, Duarte AC et al (2021) Microplastics in landfill leachates: the need for reconnaissance studies and remediation technologies. Case Stud Chem Environ Eng 3. Elsevier:100072
Somani M, Datta M, Ramana G et al (2018) Investigations on fine fraction of aged municipal solid waste recovered through landfill mining: case study of three dumpsites from India. Waste Manag Res 36(8) SAGE Publications Ltd:744–755

Somani M, Datta M, Ramana G et al (2020) Contaminants in soil-like material recovered by landfill mining from five old dumps in India. Process Saf Environ Prot 137:82–92
Somani M, Datta M, Ramana GV et al (2022) Effect of depth of landfill on the characteristics of soil-like material of aged waste: a case study of Bhalswa dumpsite, India. J Mater Cycles Waste Manag 24(5) Springer:1902–1912
Sorlini S, Abba A, Collivignarelli C (2011) Recovery of MSWI and soil washing residues as concrete aggregates. Waste Manag 31(2):289–297
Su Y, Zhang Z, Wu D, et al. (2019) Occurrence of microplastics in landfill systems and their fate with landfill age. Water Res 164. Elsevier: 114968
UNEP (2024) Beyond an age of waste—turning rubbish into a resource. Glob Waste Manag Outlook 2024. Available at: https://wedocs.unep.org/20.500.11822/44939
van Praagh M, Hartman C, Brandmyr E (2018) Microplastics in landfill leachates in the nordic countries. Nordisk Ministerråd. https://doi.org/10.6027/TN2018-557
Wanka S, Münnich K, Fricke K (2017) Landfill mining-wet mechanical treatment of fine MSW with a wet jigger. Waste Manag 59(3):316–323
Wu X, Shen D, Hui C, et al. (2024) Evaluation of pathogen spread risk from excavated landfill. Environ Pollut 349. Elsevier: 123993
Xu Z, Sui Q, Li A et al (2020) How to detect small microplastics (20–100 μm) in freshwater, municipal wastewaters and landfill leachates? A trial from sampling to identification. Sci Total Environ 733. Elsevier:139218
Zari M, Smith R, Wright C, et al. (2022) Health and environmental impact assessment of landfill mining activities: a case study in Norfolk, UK. Heliyon 8(11). Elsevier

Chapter 9
Microplastics in Gravel and Soil-like Fraction Obtained from Landfill Mining

Deepak Kumar Haritwal, G. V. Ramana, and Manoj Datta

9.1 Introduction

Plastics have become an essential component of modern society due to their affordability, versatility, and diverse applications, with their manufacture growing from 390.7 million tonnes in 2021 to 400 million tonnes in 2022 (Plastics Europe 2024). However, Poor waste management, insufficient legislation, and a lack of public awareness have contributed to severe environmental issues, including the escalating problem of MP pollution (Samandra et al. 2022, 2023). Additionally, rapid urbanisation, population growth, and changing lifestyles have significantly increased the production of Municipal Solid Waste (MSW), much of which includes plastics (Haritwal et al. 2024a). Inefficient recycling and disposal practices have exacerbated this issue, with 21% to 42% of plastic waste ending up in landfills or dumpsites (Geyer et al. 2017). With time, these plastics degrade into MPs, posing substantial environmental risks if waste is disposed of in open dumpsites without liner protection. Plastic waste accumulation and MP formation in these sites intensify pollution, threatening terrestrial environments and underscoring the pressing need for improved waste

D. K. Haritwal (✉) · G. V. Ramana · M. Datta
Department of Civil Engineering, Indian Institute of Technology Delhi, New Delhi, India
e-mail: deepak.Kumar.Haritwal@civil.iitd.ac.in

M. Somani et al. (eds.), *Landfill Mining of Legacy MSW: Global Insights and Indian Experience*, Springer Transactions in Civil and Environmental Engineering, https://doi.org/10.1007/978-981-95-3959-8_9

management practices (Haritwal et al. 2023, 2024a). One promising approach is landfill mining (LFM), which involves excavating and segregating old MSW into different fractions, primarily focusing on recovering a fine Landfill-Mined-Soil-Like-Material (LMSLM). The potential application of LMSLM necessitates a comprehensive evaluation of its contaminant profile, as studies have reported significant levels of trace metals, organic pollutants, and soluble salts (Orupõld et al. 2022; Somani et al. 2023). While the reuse potential of obtained LMSLM has been explored in various applications, such as earth-filling, subbase materials, landfill covers, compost, and agriculture, research has predominantly focused on trace metals, organic pollutants and soluble salts, overlooking other potential contaminants, especially MPs (Somani et al. 2020). This limited scope highlights the need for a more holistic evaluation to ensure safe and sustainable reuse.

The application of LMSLM in roads, highways, and other large-scale infrastructure projects across India raises significant concerns due to the lack of prolonged studies on MP migration from such materials. Without precise regulation for adequately handling and applying LMSLM, the potential for environmental impacts remains significant. Studies have identified MP presence in different landfill compartments, including landfill leachate, landfill refuse, landfill soils, and nearby groundwater (GodvinSharmila et al. 2023; He et al. 2019; Mohammadi et al. 2022). Table 9.1 highlights the MP study in landfill/dumpsite regions, revealing its high concentration. These regions exhibit significantly higher concentrations of MPs compared with agricultural soils and natural sediments (Ding et al. 2020; Ranjani et al. 2021). The high prevalence of MPs suggests their likely presence in LMSLM, warranting immediate attention to their potential impacts and underscoring the urgent need for effective waste management to reduce ecological risks (Guo et al. 2024; Wang et al. 2021).

To address these critical gaps, the study first (i) examine the abundance of MPs in mined materials (LMSLM) extracted from landfill mining activities, (ii) analyse the physiochemical characteristics of MPs, and (iii) conduct a small-scale field study to evaluate the long-term migration of MPs from LMSLM into the underlying soil and groundwater in areas where this material is used for geotechnical applications. The outcomes of this study offer valuable insights for shaping policies regarding the application of mined material in different geotechnical applications.

Table 9.1 MPs study in landfill/dumpsite regions highlighting its high concentration

Sr. No	Locations	Type of samples	MPs concentration (items/kg)	References
1	Okhla	LMSLM (India)	28,183 ± 2154	Present study
	Ghazipur,		29,850 ± 3453	
	Bhalswa		38,883 ± 1966	
2	China	Landfill refuse	62,000 ± 23,000	Su et al. (2019a, b)
3	China	Landfill refuse	50,368 ± 40,813	Wan et al. (2022)
4	Bangladesh	Landfill soil	220	Afrin et al. (2020)
5	Thailand	Landfill soil	1458	Puthcharoen and Leungprasert (2019)
6	China	Landfill soil	18,760	Zhang and Liu (2018)
7	India	Mined material	31,911 ± 2159	Goli and Singh (2023)
8	China	Landfill refuse	49,000 ± 21,000	Zhang et al. (2021)
9	India	E-waste	3519 ± 4382	Tun et al. (2022)
10	Philippines	Landfill soils	20,333 ± 12,897	
11	Vietnam	Recycling facility	20,608 ± 23,633	
12	Laos	Garbage incineration	8402 ± 7872	
13	Indonesia	Dumpsite soils	10,929 ± 13,547	
14	Cambodia	Garbage incineration	29,749 ± 67,488	

9.2 Site Description

This study examines three operational MSW dumpsites in New Delhi, India, which have operated since 1990. As shown in Fig. 9.1, these sites have collectively amassed around 28 million tonnes of MSW. The total area covered by these landfills is 200 acres, with waste predominantly mixed in nature, as reported by the operators. New Delhi produces around 12,000 metric tonnes of MSW daily, posing considerable challenges for handling such huge waste. In accordance with directives from NGT India, landfill mining activities are currently being carried out at all three dumpsites. The segregation process categorises legacy waste into three size fractions: (i) less than 6 mm, (ii) 6–30 mm, and (iii) greater than 30 mm. As shown in Fig. 9.2, the trommels used at these sites are depicted. This study concentrates on the fraction of MSW less than 30 mm, known as LMSLM, which is widely utilised for various geotechnical applications. Three sets of samples were gathered from each dumpsite, with a total of nine bulk samples for MP analysis. MSW fraction exceeding 30 mm, primarily consisting of construction and demolition debris, is excluded from this study.

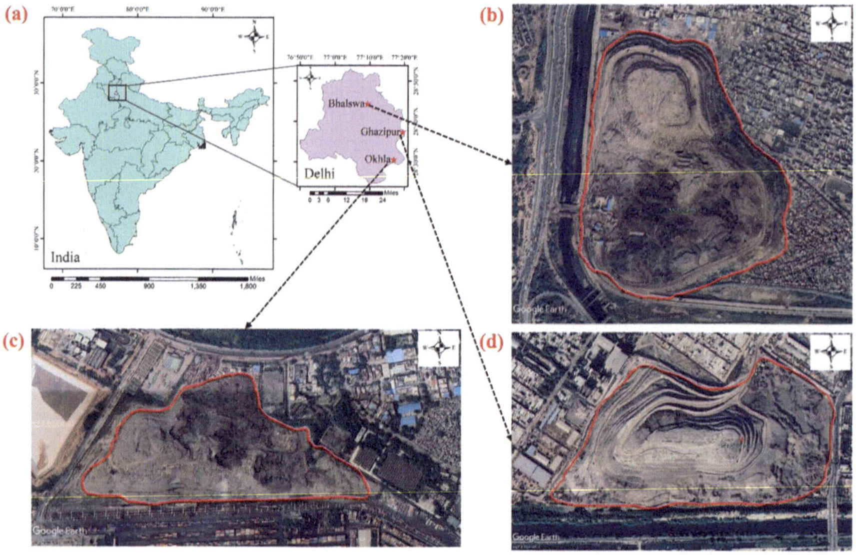

Fig. 9.1 (**a**) Indian Map highlighting dumpsite locations in New Delhi, (**b**) Bhalswa, (**c**) Okhla, and (**d**) Ghazipur dumpsite

Fig. 9.2. (**a** and **b**) Trommels used for MSW segregation into different fractions, (**c**) >30 mm, (**d**) <30 mm (LMSLM), and (**e**) <6 mm

9.3 Extraction and Characterisation of MPs

Samples were first air-dried to minimise surface moisture and then oven-dried at 40 °C. This drying process was crucial for addressing moisture-related issues, as excess moisture could cause fines to agglomerate, clogging the sieves and hindering accurate MP extraction (Haritwal et al. 2024b). Moisture can cause MPs to adhere to the soil surface, potentially leading to their loss during extraction. Once dried, the material was sieved using a 4.75 mm mesh to ensure uniformity and prepare them for further analysis. Approximately 20 g of samples were moved to a glass beaker, and the organic matter was removed using the Fenton Reagent. The solution was poured gradually until the removal of foam at the top surface, and after that, the solution was then placed in an oven operating at 40 °C for 24 hours to reduce its volume without altering the morphology of the polymers. After organic matter digestion, a dual-density (NaCl & $ZnCl_2$) separation method was applied to enable separation based on density. The solution was left to settle overnight and later passed by a 0.45 µm pore size filter paper. The extracted MPs were visually seen using a stereo microscope. The MPs' concentration, shape, and

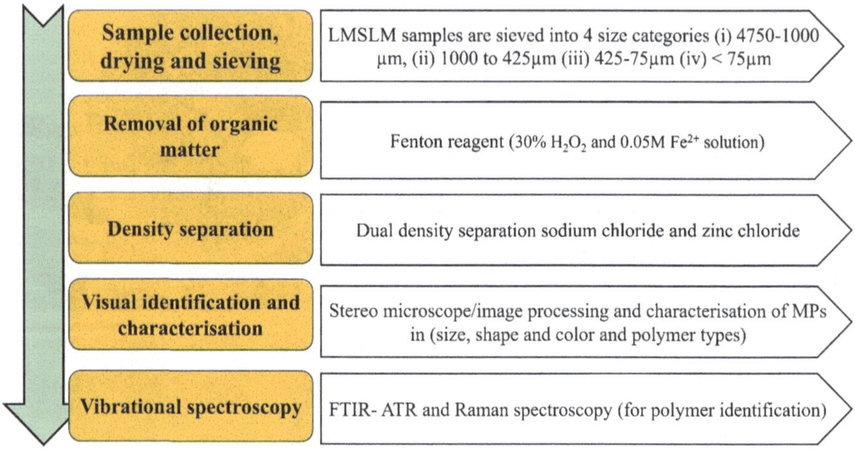

Fig. 9.3 Flow chart highlights the steps used for MPs extraction and characterisation from LMSLM

colour were manually documented, while imageJ was used to classify MP size. Polymer identification was conducted FTIR-ATR and Raman Spectroscopy. The MP surface morphology and elemental composition were also analysed using SEM-EDX. Figure 9.3 illustrates the complete process for MP extraction and characterisation from LMSLM.

9.4 Results and Discussions

9.4.1 Concentration of MPs

MPs were found in all samples with concentrations (items/kg) varying with dumpsites, as shown in Fig. 9.4. The average concentration of MPs was 28,183 ± 2154 items/kg at the Okhla dumpsite, 29,850 ± 3453 items/kg at the Ghazipur dumpsite, and 38,883 ± 1966 items/kg at the Bhalswa dumpsite. The lowest concentration of MPs, 25,950 items/kg, was observed at the Okhla dumpsite, while the highest concentration, 41,100 items/kg, was found at the Bhalswa dumpsite. The historical deposition patterns, waste composition, and the heterogeneous nature of MSW at these sites likely influence this variability. The MP concentrations detected in the current study were significantly higher than those documented in agricultural soils and soil sediments (Castan et al. 2021; Corradini et al. 2019). However, these values were lower than those found in landfill refuse. This underscores the persistence of plastic and emphasises the influence of environmental parameters such as temperature, weather, and physical abrasion in driving MP generation over time (Golwala et al. 2021; Su et al. 2019b). The variation in concentrations observed in the current study reflects global trends, where waste management practices shape waste composition and deposition volumes. Regions such as the USA, Canada, and Western

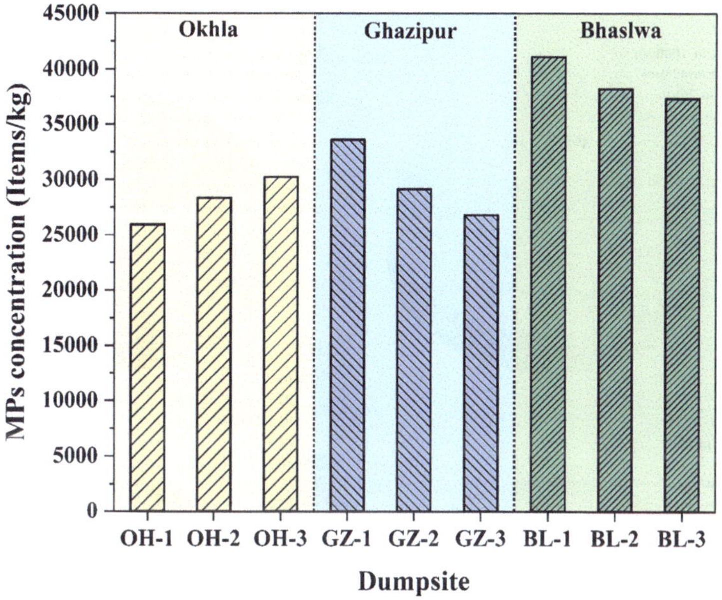

Fig. 9.4 MPs concentration in LMSLM across dumpsites

Europe typically produce waste streams with higher proportions of packaging materials and plastics. In contrast, India has experienced a significant increase in plastic loads at dumpsites due to rapid urbanisation, elevated demand for packaged items, and contributions from household refuse, industrial waste materials, and electronic waste (Sun et al. 2021; Wang et al. 2021). This study emphasises the regional factors in determining MP concentrations in LMSLM. Although reusing LMSLM offers a viable solution for managing large waste volumes, particularly in developing countries like India, its unprotected application in geotechnical projects could lead to MP migration into subsurface soils and groundwater, causing long-term ecological damage. Results highlight the rapid need for region-specific waste management approaches and rigorous assessment protocols to curb MP pollution and ensure the safe reuse of LMSLM in large-scale applications.

9.4.2 MPs Characteristics

The collected MPs were classified based on size, shape, colour, and polymer type. The study divided MPs into four sizes according to their maximum length, as shown in Fig. 9.5(a). Most MPs fell within the 425–75 μm range, accounting for 52%. MPs in the 1000–425 μm range constituted 29%, while those in the 5000–1000 μm range made up 11%. Notably, MPs smaller than 75 μm, representing 8%, were detected across all LMSLM samples. The consistent presence of small MPs highlights their

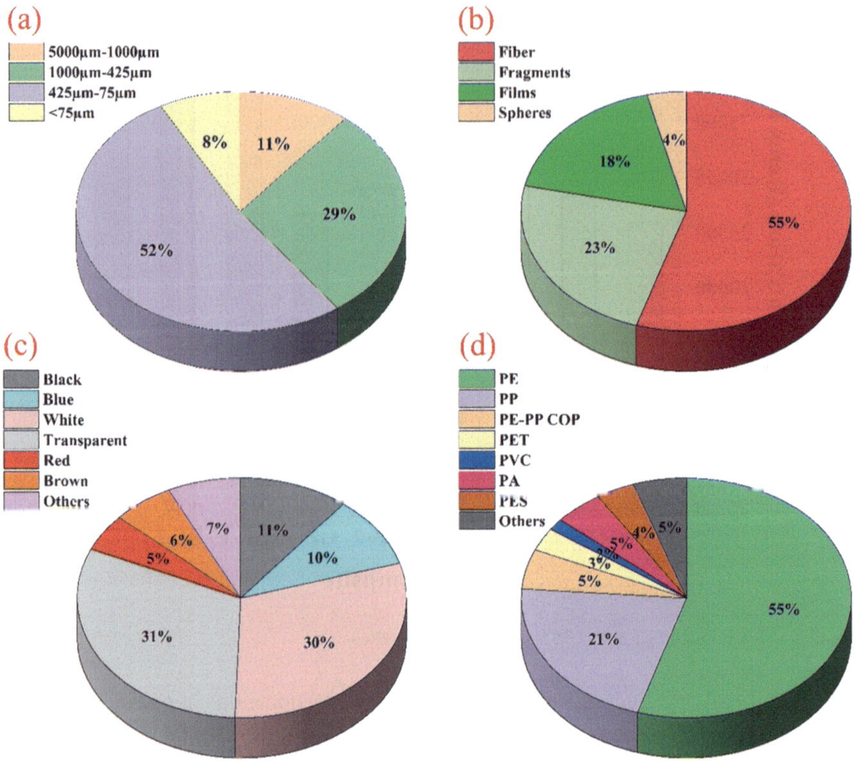

Fig. 9.5 MPs classification based on (**a**) Size, (**b**) Shape, (**c**) Colour, and (**d**) Polymer types

widespread prevalence. At the same time, minor fluctuations in MP fractions may be due to the heterogeneous distribution of MSW over time (Sun et al. 2021). Figure 9.6 highlights the size fraction of MPs observed within LMSLM. The results are aligned with studies reported by Manikanda et al. (2021) and Uddin et al. (2021), which emphasise that landfill age is a critical factor responsible for the concentration of smaller MP fractions. Additionally, factors such as seasonal changes, environmental weathering and complex biochemical reactions may accelerate MP degradation (Bandow et al. 2017; Li et al. 2024). Application of LMSLM in areas with shallow groundwater tables is vulnerable to groundwater pollution because of small MPs. Furthermore, these sites, which have been functioning for more than 40 years, demonstrate the influence of weathering and seasonal variations on the presence of small-sized MPs. Considering the documented health risks associated with smaller MPs, it is vital to fully evaluate the size distribution of MPs in LMSLM before considering any reuse applications.

MPs were categorised into four morphological types: As shown in Fig. 9.5b, fibres (55%) were the predominant MP shape across all samples, followed by fragments (23%), films (18%), and spheres (4%). These findings align with previous studies that document fibres shape MPs dominance in landfill environments

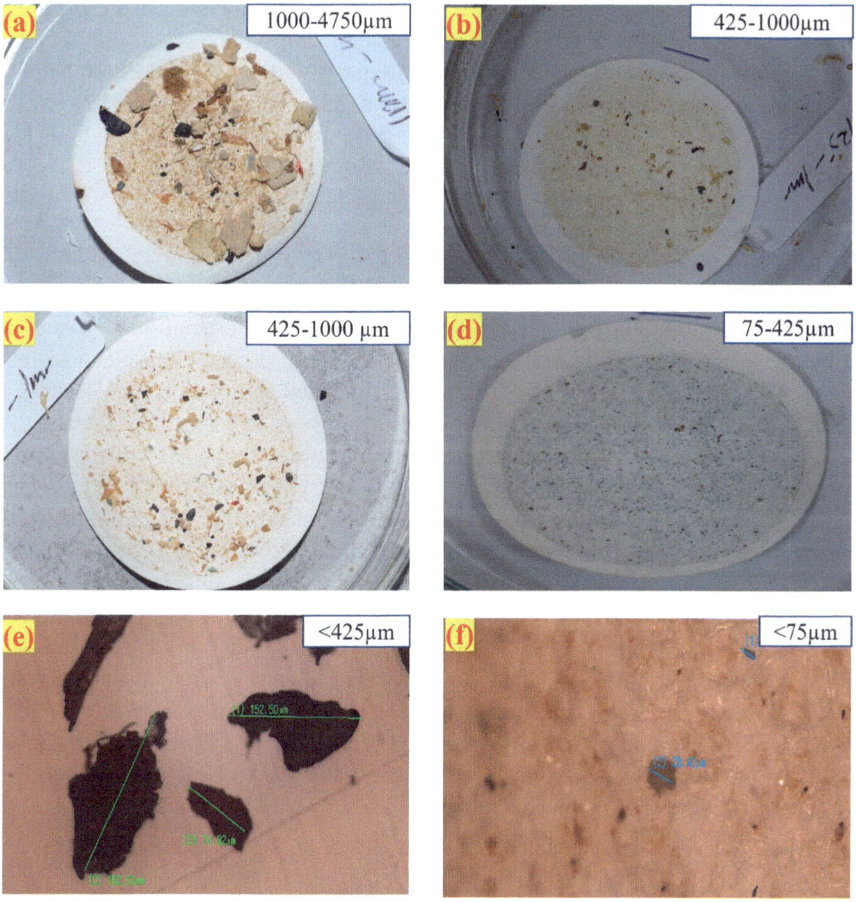

Fig. 9.6 MPs observed in different size fractions of LMSLM

(Manikanda et al. 2021; Panno et al. 2019). MP distribution based on colour and polymer type is shown in Fig. 9.5c,d. The most common MP colours identified were white (30%), transparent (31%), black (11%), blue (10%), and brown (6%). The presence of white and transparent MPs aligns with their widespread use of packaging. Similar trends have been studied by (Kazour et al. 2019). The observed colour of MPs corresponds to the parent colours of the plastic disposed of in MSW, although various environmental processes can alter these colours over time. Furthermore, previous research has noted a high quantity of white/transparent MPs in different matrices, including soils, sediments and water bodies. As shown in Fig. 9.5d, the results underscore the presence of different polymer types, with Polyethylene (PE) being the most prevalent at (55%), followed by Polypropylene (PP) at (21%), PE-PP Copolymer at (5%), Polyethylene Terephthalate (PET) at (3%), Polyamide (PA) and Polyester (PES) each at (4%), and Polyvinyl Chloride

(PVC) at (2%). The dominance of PE and PP highlights their widespread global production and extensive application in the manufacturing and packaging sectors. These findings align with other studies that frequently report PE and PP as the predominant polymers in waste sites (Golwala et al. 2021; Johar et al. 2023).

9.4.3 Sem-EDX Analysis of MPs

Surface morphology of MPs was examined using SEM. The analysis indicated that UV radiation, high temperature, and mechanical stress are crucial influences contributing to the weathering and ageing of MPs. These processes result in visible signs of deterioration, including pores, pits and attached particles on the MP surface (Cai et al. 2018). SEM images highlighted differences in surface roughness; the rough texture was observed with fragments while fibres shape MPs display comparatively smoother surface, as shown in Fig. 9.7a, b cracks and rough surface, and (c and d) attached particles on MPs surface. As illustrated in Fig. 9.7e, f, the analysis identified several elements on the MP surface, including C, O, Fe, Ca, Si, Pb, As, Hg, Ti, Cd, and Zn. C and O were the most abundant, while notable levels of trace metals, including Cd, Pb, and As, were detected. These trace metals are deliberately incorporated during plastic manufacturing to improve their durability; for example, silica is added to improve physical durability, and Al and Si serve as antioxidants (Nematollahi et al. 2020). Furthermore, these trace metals may adsorb onto MPs from the surrounding area. The results indicate that MPs can adsorb these trace metals on their surface, which could pose serious risks due to their leakage into the terrestrial environment.

9.4.4 Field Study

A field study was conducted near the Bhalaswa dumpsite to measure the migration of MPs from LMSLM to the underlying soil and groundwater. Historically, this material has been dispersed in open areas near the site. This region was historically filled with mined material as part of restoration projects initiated under cleaner initiatives mandated by India's National Green Tribunal (NGT). The study employed advanced field techniques to find the spread of MPs in the subsurface. Continuous samples were retrieved via dual-tube sampling systems, which ensured minimal disturbance and preserved sample integrity in well-packed liners, as shown in Fig. 9.8 (Singh et al. 2024, 2025). Groundwater samples were subsequently extracted from high-permeability zones, enabling an assessment of MP contamination in both soil and water matrices. The methodology ensured high-quality sample collection and provided valuable data on the extent of MP infiltration from the unsaturated zone to the saturated zone. MP concentrations were measured at 15 cm intervals from the SLM/LMSLM- interface to groundwater, offering a detailed

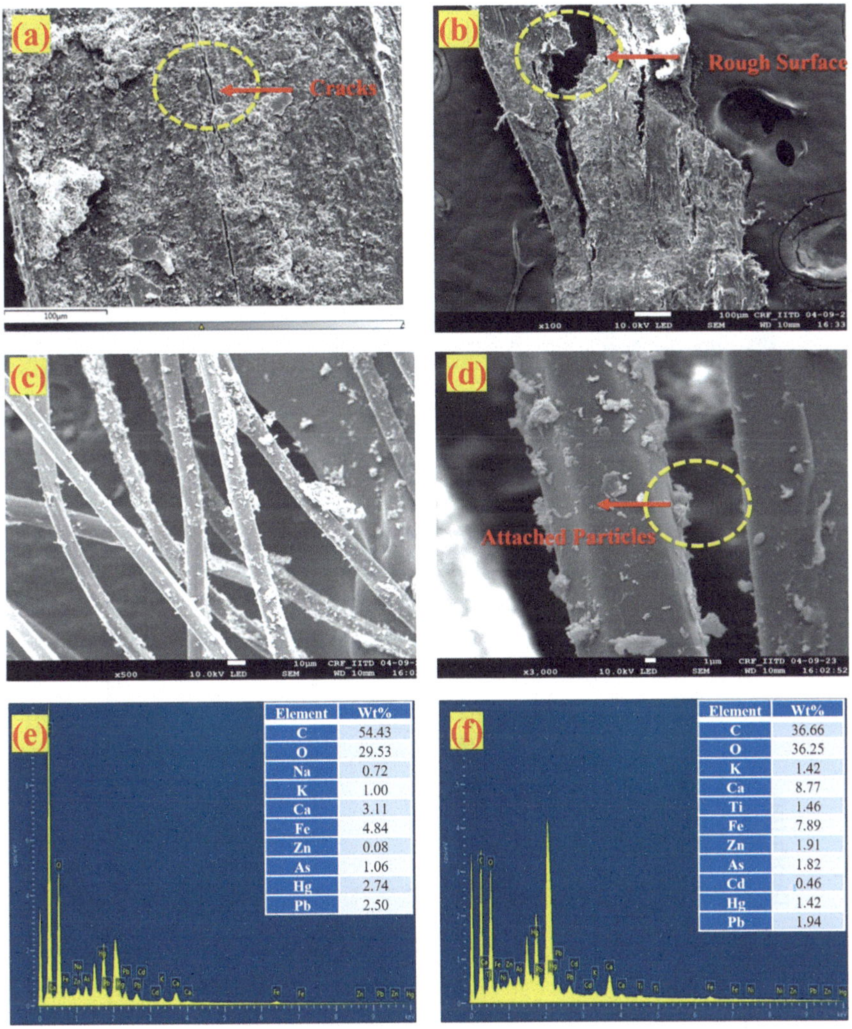

Fig. 9.7 SEM-EDX analysis of observed MPs in LMSLM

vertical contamination profile. Groundwater was collected from multiple depths at the same location to understand the extent of MP leaching into aquifers. This field study is significant in assessing the prolonged environmental impact posed by LMSLM, particularly in regions with high groundwater vulnerability. By integrating advanced sampling techniques, the research reveals the likely possibility for MPs to migrate over time, driven by environmental and hydrological processes, thereby underscoring the need to manage landfill materials carefully to mitigate long-term contamination risks. Figure 9.8 depicts the site location and sampling points, and the variation of MP concentration with depth is shown in Fig. 9.9.

Fig. 9.8 Field study location (**a**) near Bhalswa dumpsite, (**b**) SP4 and SP5 area filled with LMSLM obtained from mining activities, (**c**) Extracted continuous soil samples underlying LMSLM materials

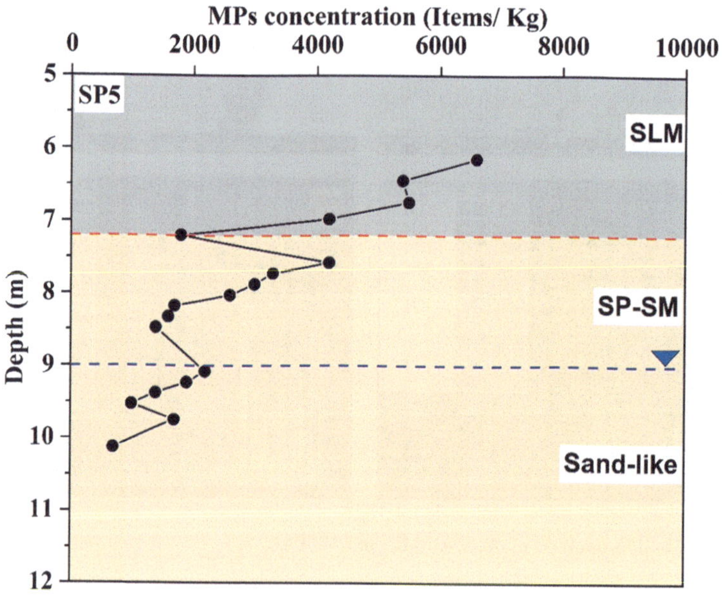

Fig. 9.9 MPs concentration profile with depth

Results revealed an MP concentration of 11,700 ± 3456 items/kg at the top surface, with slight fluctuations recorded within the LMSLM zone. These concentrations gradually decreased to 7400 ± 1800 items/kg near the LMSLM-underlying soil (SLM-SM) interface. A consistent downward trend was noted with depth, with concentrations reducing by 80–95%, reaching 1000 ± 20 items/kg in groundwater, as depicted in Fig. 9.9. The findings highlight the role of subsurface soil in restricting MP migration. Interestingly, an increase in MPs was detected at the waste/soil border, likely attributed to variations in soil structure and drainage characteristics. The findings underscore concentrations of MPs of (171) and (105) items/L) at these points in the groundwater. Despite these high concentrations, they are frequently used. However, limited information is available for the prolonged implications of these materials on their potential to pollute groundwater. These results emphasise the need to comprehensively assess such practices and develop appropriate measures to mitigate risks.

9.5 Conclusions

- Results revealed high concentrations of MPs in LMSLM, where landfill mining is ongoing. MP concentrations varied significantly, with 28,183 ± 2154 items/kg at Okhla, 29,850 ± 3453 items/kg at Ghazipur, and 38,883 ± 1966 items/kg at Bhalswa dumpsite. A significant portion of MPs (60–70%) were smaller than 425 µm, with 5–10% under 75 µm. Fibres were the most common shapes (50–60%). Predominantly white and transparent MPs, mainly PE and PP, with trace amounts of PA, PET, and PES were observed. These findings underscore the ecological risks due to the direct application of LMSLM in environmental projects, as MP's small size and mobility pose contamination threats to nearby ecosystems.
- The field study revealed high concentrations of MPs just above the interface of the soil-like material (SLM/LMSLM) and underlying soil, with values reaching 7400 items/kg, while concentrations in the saturated zone at the bottom were lower and decreased with depth, indicating continuous migration. MPs were also detected in groundwater samples from high-permeability zones, with concentrations up to 171 items/L, decreasing with depth. The results showed a predominance of small, fibre-shaped MPs, like soil samples, emphasising the dumpsite's impact on groundwater quality.
- The finding emphasises the importance of developing reuse strategies that can help in reducing the risks due to MPs to prevent further environmental contamination. A practical solution could effectively reduce MP migration by using a low-permeable clay layer. This study offers valuable information on MP contamination, paving the way for better practices and promoting more sustainable strategies for LMSLM reuse.

References

Afrin S, Uddin MK, Rahman MM (2020) Microplastics contamination in the soil from urban landfill site, Dhaka, Bangladesh. Heliyon 6(11). https://doi.org/10.1016/j.heliyon.2020.e05572

Bandow N, Will V, Wachtendorf V, Simon FG (2017) Contaminant release from aged microplastic. Environ Chem 14(6):394–405. https://doi.org/10.1071/EN17064

Cai L, Wang J, Peng J, Wu Z, Tan X (2018) Observation of the degradation of three types of plastic pellets exposed to UV irradiation in three different environments. Sci Total Environ 628–629:740–747. https://doi.org/10.1016/j.scitotenv.2018.02.079

Castan S, Henkel C, Hüffer T, Hofmann T (2021) Microplastics and nanoplastics barely enhance contaminant mobility in agricultural soils. Commun Earth Environ 2(1). https://doi.org/10.1038/s43247-021-00267-8

Corradini F, Meza P, Eguiluz R, Casado F, Huerta-Lwanga E, Geissen V (2019) Evidence of microplastic accumulation in agricultural soils from sewage sludge disposal. Sci Total Environ 671:411–420. https://doi.org/10.1016/j.scitotenv.2019.03.368

Ding L, Zhang S, Wang X, Yang X, Zhang C, Qi Y, Guo X (2020) The occurrence and distribution characteristics of microplastics in the agricultural soils of Shaanxi Province, in North-Western China. Sci Total Environ 720. https://doi.org/10.1016/j.scitotenv.2020.137525

Geyer R, Jambeck JR, Law KL (2017) Production, use, and fate of all plastics ever made. https://www.science.org. Sci Adv 3

GodvinSharmila V, Shanmugavel SP, Tyagi VK, Rajesh Banu J (2023) Microplastics as emergent contaminants in landfill leachate: source, potential impact and remediation technologies. J Environ Manag 343. https://doi.org/10.1016/j.jenvman.2023.118240

Goli VSNS, Singh DN (2023) Extraction and characterisation of microplastics in landfill-mined-soil-like-fractions: a novel methodology. Chem Eng J 452. https://doi.org/10.1016/j.cej.2022.139217

Golwala H, Zhang X, Iskander SM, Smith AL (2021) Solid waste: an overlooked source of microplastics to the environment. Sci Total Environ 769). Elsevier B.V. https://doi.org/10.1016/j.scitotenv.2020.144581

Guo S, Wu Z, Li X, Shen D, Shentu J, Lu L, Qi S, Zhu M, Long Y (2024) Microplastic, a possible trigger of landfill sulfate reduction process. Sci Total Environ 906:167662. https://doi.org/10.1016/J.SCITOTENV.2023.167662

Haritwal D, Ramana GV, Datta M (2023) Microplastics in soil-like-material obtained during mining of a municipal solid waste dumpsite. 25–28. https://doi.org/10.53243/ICEG2023-129

Haritwal DK, Singh P, Ramana GV, Datta M (2024a) Advanced characterisation of groundwater contamination at a dumpsite: methodology and assessment—case study of a municipal solid waste dumpsite in India. Sci Total Environ 957. https://doi.org/10.1016/j.scitotenv.2024.177642

Haritwal DK, Singh P, Ramana GV, Datta M (2024b) Microplastic migration from landfill-mined soil through earth filling operations and ecological risk assessment: a case study in New Delhi, India. Environ Sci Pollut Res Int. https://doi.org/10.1007/s11356-024-35545-3

He P, Chen L, Shao L, Zhang H, Lü F (2019) Municipal solid waste (MSW) landfill: a source of microplastics?-evidence of microplastics in landfill leachate. Water Res 159:38–45. https://doi.org/10.1016/j.watres.2019.04.060

Johar IS, Archana AR, Gopika B, Mina M, Haddout S, Madhu A (2023) Implications of solid waste dumps on the microplastic abundance in groundwater in Kollam, India. J Environ Manag 348. https://doi.org/10.1016/j.jenvman.2023.119224

Kazour M, Terki S, Rabhi K, Jemaa S, Khalaf G, Amara R (2019) Sources of microplastics pollution in the marine environment: importance of wastewater treatment plant and coastal landfill. Mar Pollut Bull 146:608–618. https://doi.org/10.1016/j.marpolbul.2019.06.066

Li X, Guo S, Shen D, Shentu J, Lv L, Qi S, Zhu M, Long Y (2024) Microplastic release and sulfate reduction response in the early stage of a simulated landfill. Waste Manag 175:22–29. https://doi.org/10.1016/J.WASMAN.2023.12.037

Manikanda BK, Natesan U, Vaikunth R, Praveen K, Ruthra R, Srinivasalu S (2021) Spatial distribution of microplastic concentration around landfill sites and its potential risk on groundwater. Chemosphere 277:130263. https://doi.org/10.1016/j.chemosphere.2021.130263

Mohammadi A, Malakootian M, Dobaradaran S, Hashemi M, Jaafarzadeh N (2022) Occurrence, seasonal distribution, and ecological risk assessment of microplastics and phthalate esters in leachates of a landfill site located near the marine environment: Bushehr port, Iran as a case. Sci Total Environ 842. https://doi.org/10.1016/j.scitotenv.2022.156838

Nematollahi MJ, Moore F, Keshavarzi B, Vogt RD, Nasrollahzadeh Saravi H, Busquets R (2020) Microplastic particles in sediments and waters, south of Caspian Sea: frequency, distribution, characteristics, and chemical composition. Ecotoxicol Environ Saf 206. https://doi.org/10.1016/j.ecoenv.2020.111137

Orupõld K, Somani M, Kaczala F, Hogland M, Bhatnagar A, Hogland W, Saaremäe E, Kriipsalu M (2022) Ecotoxicity assessment of fine fractions obtained from landfill mining. J Hazard Toxic Radioact Waste 26(4). https://doi.org/10.1061/(asce)hz.2153-5515.0000715

Panno SV, Kelly WR, Scott J, Zheng W, McNeish RE, Holm N, Hoellein TJ, Baranski EL (2019) Microplastic contamination in karst groundwater systems. Ground Water 57(2):189–196. https://doi.org/10.1111/gwat.12862

Plastics Europe, 2024 (2024). https://doi.org/https://plasticseurope.org/knowledge-hub/the-circular-economy-for-plastics-a-european-analysis-2024/

Puthcharoen A, Leungprasert S (2019) Determination of microplastics in soil and leachate from the landfills. Thai Environ Eng J 33(3) www.eeat.or.th

Ranjani M, Veerasingam S, Venkatachalapathy R, Mugilarasan M, Bagaev A, Mukhanov V, Vethamony P (2021) Assessment of potential ecological risk of microplastics in the coastal sediments of India: a meta-analysis. Mar Pollut Bull 163. https://doi.org/10.1016/j.marpolbul.2021.111969

Samandra S, Johnston JM, Jaeger JE, Symons B, Xie S, Currell M, Ellis AV, Clarke BO (2022) Microplastic contamination of an unconfined groundwater aquifer in Victoria, Australia. Sci Total Environ 802. https://doi.org/10.1016/j.scitotenv.2021.149727

Samandra S, Singh J, Plaisted K, Mescall OJ, Symons B, Xie S, Ellis AV, Clarke BO (2023) Quantifying environmental emissions of microplastics from urban rivers in Melbourne, Australia. Mar Pollut Bull 189. https://doi.org/10.1016/j.marpolbul.2023.114709

Singh P, Haritwal DK, Seth S, Ramana GV, Datta M (2024) Electrical conductivity profiling for rapid contamination assessment in unsaturated zones: a case study of an MSW landfill. Sci Total Environ 951. https://doi.org/10.1016/j.scitotenv.2024.175773

Singh P, Haritwal DK, Ramana GV, Datta M (2025) Framework for site specific leachate limits in MSW dumpsites: integration of high-resolution site characterisation and contaminant transport modeling. Chemosphere 371. https://doi.org/10.1016/j.chemosphere.2024.144054

Somani M, Datta M, Ramana GV, Sreekrishnan TR (2020) Contaminants in soil-like material recovered by landfill mining from five old dumps in India. Process Saf Environ Prot 137:82–92. https://doi.org/10.1016/j.psep.2020.02.010

Somani M, Hölzle I, Datta M, Ramana GV (2023) An investigation on mobility of heavy metals for assessing the reusability of soil-like material reclaimed from mining of municipal solid waste dumpsites. Waste Manag 167:113–121. https://doi.org/10.1016/j.wasman.2023.05.028

Su Y, Zhang Z, Wu D, Zhan L, Shi H, Xie B (2019a) Occurrence of microplastics in landfill systems and their fate with landfill age. Water Res 164:114968. https://doi.org/10.1016/J.WATRES.2019.114968

Su Y, Zhang Z, Wu D, Zhan L, Shi H, Xie B (2019b) Occurrence of microplastics in landfill systems and their fate with landfill age. Water Res 164. https://doi.org/10.1016/j.watres.2019.114968

Sun J, Zhu ZR, Li WH, Yan X, Wang LK, Zhang L, Jin J, Dai X, Ni BJ (2021) Revisiting microplastics in landfill leachate: unnoticed tiny microplastics and their fate in treatment works. Water Res 190. https://doi.org/10.1016/j.watres.2020.116784

Tun TZ, Kunisue T, Tanabe S, Prudente M, Subramanian A, Sudaryanto A, Viet PH, Nakata H (2022) Microplastics in dumping site soils from six Asian countries as a source of plastic additives. Sci Total Environ 806. https://doi.org/10.1016/J.SCITOTENV.2021.150912

Uddin S, Fowler SW, Uddin MF, Behbehani M, Naji A (2021) A review of microplastic distribution in sediment profiles. Mar Pollut Bull 163. https://doi.org/10.1016/j.marpolbul.2021.111973

Wan Y, Chen X, Liu Q, Hu H, Wu C, Xue Q (2022) Informal landfill contributes to the pollution of microplastics in the surrounding environment. Environ Pollut 293. https://doi.org/10.1016/j.envpol.2021.118586

Wang Y, Lu X, Fei X (2021) Property changes of conventional plastic waste mixed with municipal solid waste after 10-year degradation experiments simulating landfill conditions. J Hazard Mater Lett 2. https://doi.org/10.1016/j.hazl.2021.100047

Zhang GS, Liu YF (2018) The distribution of microplastics in soil aggregate fractions in southwestern China. Sci Total Environ 642:12–20. https://doi.org/10.1016/j.scitotenv.2018.06.004

Zhang Y, Peng Y, Peng C, Wang P, Lu Y, He X, Wang L (2021) Comparison of detection methods of microplastics in landfill mineralized refuse and selection of degradation degree indexes. Environ Sci Technol 55(20):13802–13811. https://doi.org/10.1021/acs.est.1c02772

Chapter 10
Characterization and Valorization Potential of Landfill-Derived Soil (−4.75 mm): Physico-chemical Insights and Pollution Risks Across Four Cities

Ajay S. Kalamdhad, Vaishnavi Jahagirdar, Arif Khan, and Anil Kumar Mishra

10.1 Introduction

Many developing countries face significant challenges in managing substantial quantities of municipal solid waste (MSW) accumulated over time through open dumping, often with little attention to the generation of leachate and methane emissions (Kaushal et al. 2012). Enhanced landfill mining, an emerging approach in these regions, involves the systematic excavation, screening, sorting, segregation and treatment of legacy waste to recover value-added products. These products can substitute primary resources, thereby promoting the transition to a circular economy (Jones et al. 2013). Beyond resource recovery and circular economy benefits, landfill mining also facilitates the reclamation of urban land, which can be repurposed for constructing sanitary landfills or waste management infrastructure (Joseph et al. 2008). The four primary residues generated from landfill mining include the combustible fraction, the fine fraction or soil-like material, construction and demolition (C&D) waste, and recyclables (Hogland 2002). Most studies, such as those by Jain et al. (2005), Prechthai et al. (2008), and Kaartinen et al. (2013) have focused on characterization of these residues to evaluate their potential for valorization.

The soil-like material or landfill-mined fine fraction (hereafter LMFF) is primarily composed of decomposed organic matter, cover soil, and street sweepings, along with wood chippings, fiber, plastics, and other miscellaneous materials (Zhou et al. 2015). Despite constituting approximately 45–80% of the total landfill-mined residues, studies on the valorization of LMFF remain limited due to its complex and variable characteristics, which present significant challenges for practical

A. S. Kalamdhad · V. Jahagirdar (✉) · A. Khan · A. K. Mishra
Department of Civil Engineering, Indian Institute of Technology Guwahati, Guwahati, Assam, India
e-mail: vaishnavi.prasanna@iitg.ac.in

M. Somani et al. (eds.), *Landfill Mining of Legacy MSW: Global Insights and Indian Experience*, Springer Transactions in Civil and Environmental Engineering, https://doi.org/10.1007/978-981-95-3959-8_10

applications. The study by Jahagirdar et al. (2025) highlighted the exceptionally high concentration of heavy metals in LMFF, further complicating its utilization. Existing studies have explored the potential applications of LMFF, including its use as a soil enricher by Kennedy et al. (2021), treated material for construction and earthwork by Wanka et al. (2017), a cover layer for old and new landfills (Burlakovs et al. 2021), and as a component in fired bricks (Singh and Chandel 2022). However, the feasibility of these applications depends on site-specific characteristics, particularly the heavy metal content and associated pollution risks. Additionally, the composition of legacy waste significantly influences LMFF characteristics. Due to unique waste composition and the improper practices of dumping mixed MSW, characterization studies conducted globally may not be directly applicable to Indian conditions (Singh and Chandel 2020).

This study presents a comparative analysis of LMFF procured from four different dump sites in India. The primary objective is to provide critical insights into the environmental implications of LMFF by conducting physico-chemical analyses and assessing total heavy metal concentrations, their bioavailability, and leachability. These evaluations aim to highlight the potential risks associated with heavy metals in LMFF from all four sites. Furthermore, this study is significant as it explores potential treatment strategies and off-site applications, thereby offering practical solutions for the sustainable valorization of the LMFF. The findings are particularly relevant for cities implementing enhanced landfill mining initiatives, contributing to the development of effective waste management practices.

10.2 Materials and Methodology

10.2.1 Sampling of LMFF from Different Sites

The landfill-mined fine fraction was procured from the Boragaon, Ghazipur, Bhalswa and Bandhwari dump sites as shown in Fig. 10.1., where landfill mining was underway. At the time of sampling, the trommels employed at the sites had a screening size of 30 mm at Boragaon, Ghazipur and Bhalswa and 6 mm at the Bandhwari site, resulting in the production of LMFF below these sieve sizes. Based on the observations from all the sites, LMFF was segregated into separate heaps, and representative samples were initially collected using the coning and quartering method to ensure homogeneity. These samples were then securely stored in tightly sealed containers to preserve their integrity. For experimental purposes and to ensure consistency in analysis, all collected samples were subsequently sieved through a 4.75 mm sieve to facilitate ease of comparison. Table 10.1 showcases the description of each dumpsite location.

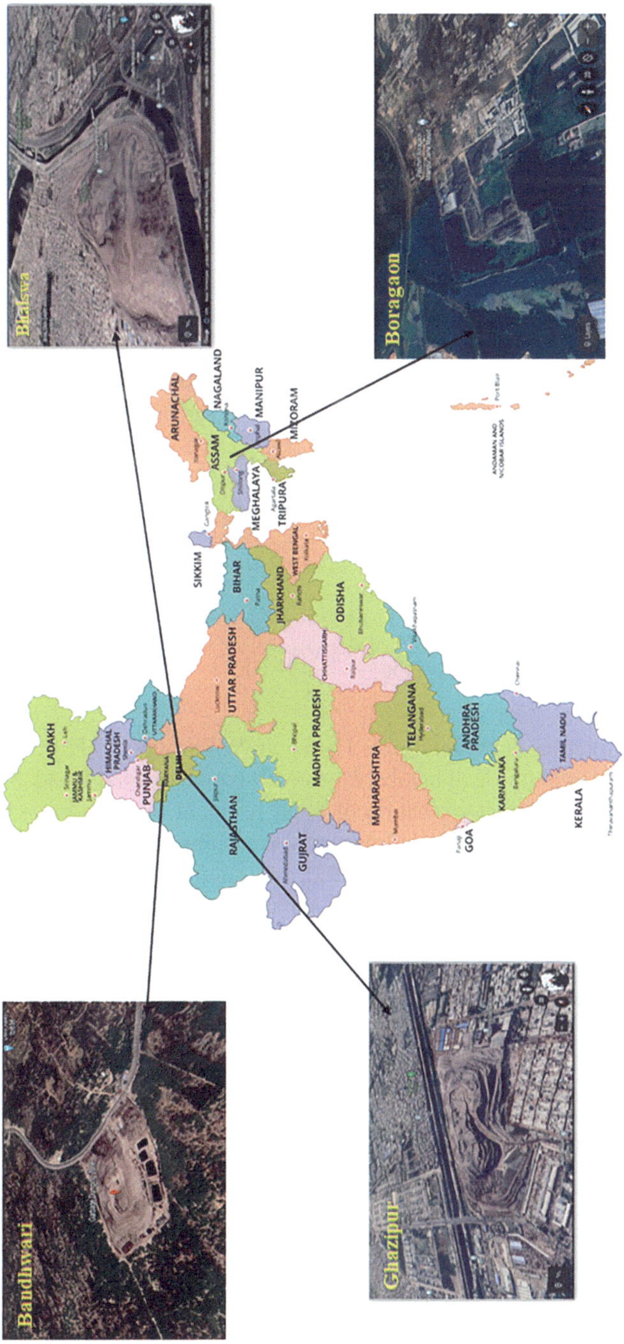

Fig. 10.1 Study area

Table 10.1 Description of locations

Sites	Boragaon	Ghazipur	Bhalswa	Bandhwari
Municipal corporation	Guwahati	East Delhi	North Delhi	Gurugram
Dumpsite operational since	2006	1984	1994	2013
Incoming waste (MT/day)	800	6500	2400	2000
Area (acres)	36	70	52	31

10.2.2 Physico-Chemical Characterization

In order to measure the pH value, dried sample passing through a 0.22 mm sieve was mixed with distilled water in liquid-to-solid (L/S) ratio of 10:1 and pH of the extracted solution was measured by a pH meter (Singh and Kalamdhad 2016). For evaluation of volatile solids (VS) which directly correlates with the organic matter, the oven dried samples were burned at 550 °C for 2 hours and % VS was calculated. The ash content was determined as follows,

$$\text{Ash content}(\%) = 100 - \text{VS} \tag{10.1}$$

The Eq. (10.2) was utilized by Reddy et al. (2018) to calculate the Total Organic Carbon (TOC).

$$\%\text{TOC} = \frac{\%\text{VS}}{1.8} \times 100 \tag{10.2}$$

In the investigation of the nutrients (Nitogen-N, Phosphorus-P, Potassium-K), the assessment of total nitrogen involved the combination of Total Kjeldahl nitrogen and nitrate content. The determination of Total Kjeldahl nitrogen was carried out using the Kjeldahl method, while Ultraviolet spectrophotometry, as outlined by the (American Public Health Association 1992), was employed for the evaluation of nitrate content. The assessment of orthophosphates was conducted utilizing molybdate spectrophotometry as outlined by (American Public Health Association 1992). Additionally, the determination of potassium and calcium employing the Inductively Coupled Plasma Mass Spectrometry (LC-ICPMS) (Make: Agilent, Model:7850 ICP-MS).

10.2.3 Total Heavy Metals, their Bioavailability and Leachability in LMFF

To ascertain the concentration of total heavy metals in the LMFF, a sample weighing 0.2 g passing through a 212 μ sieve was subjected to digestion for 2 h at a temperature of 300 °C by treating the sample with a mixture of 10 mL of H_2SO_4 and $HClO_4$ at a ratio of 5:1 (Singh and Kalamdhad 2012). The digested sample was diluted, filtered, and utilized for the total heavy metal analysis utilizing ICP-MS. To investigate the bioavailability of heavy metals, the extraction of these metals was conducted employing diethylene triamine penta-acetic acid (DTPA) as reported by (Singh and Kalamdhad 2019). DTPA solution is assumed to extract carbonate-bound and organically bound metal fractions in calcareous soils and indicates the number of metals potentially available from soils. Furthermore, the extraction process was carried out at a pH of 7.3, with an L/S ratio of 10:1 and the resulting supernatants were filtered and stored at 4 °C for heavy metal analysis. To assess the leachability of heavy metals from the LMFF, the standard toxicity characteristic leaching procedure (TCLP) as outlined by the (EPA 1992) was performed. In the extraction procedure, acetic acid was combined in L/S ratio of 20:1, precisely maintained at a pH of 4.93 ± 0.05. Subsequently, after centrifugation at 10,000 rpm, the resulting supernatant was subjected to filtration and stored at 4 °C for heavy metal analysis.

10.2.4 Pollution Indices

The determination of contamination factor (C_f^i) offers a direct means of assessing the effects of heavy metal contamination through the comparison of metal concentrations in the targeted sample (here, LMFF) against the anticipated natural or baseline values. C_f^i is quantified using the equation,

$$\text{Cfi} = \frac{\text{Heavy metal concentration in LMFF}}{\text{Heavy metal concentration in baseline}} = \frac{C_{\text{LMFF}}}{C_{\text{baseline}}} \tag{10.3}$$

Hakanson (1980) establishes four categories of contamination namely;

- $C_f^i < 1$ indicates low contamination,
- $1 \le C_f^i < 3$ indicates moderate contamination,
- $3 \le C_f^i < 6$ indicates considerable contamination, and
- $6 \le C_f^i$ indicates high contamination.

The Pollution Load Index (PLI) is a vital environmental metric used to evaluate the overall degree of contamination induced by *n* polluting elements and is computed by taking the geometric mean of the contamination factors associated with each element, as shown in the following expression given by Tomlinson et al. (1980). PLI values higher than unity suggest pollution existence, while lower than 1 indicates no pollution load.

$$\text{PLI} = \sqrt[n]{C_f^1 \times C_f^2 \times C_f^3 \times \ldots \times C_f^n} \tag{10.4}$$

The potential ecological risk index (R_I) was developed as a means of evaluating the extent of heavy metal contamination in sediments, considering their toxicity and the impact on the environment (Maanan et al. 2015). Based on this, R_I could also be evaluated for LMFF to identify the contaminants and the extent of risk it could cause for the geoenvironment. R_I is the addition of all risk factors for individual heavy metals and is evaluated as follows:

$$R_I = \sum E_r^i \tag{10.5}$$

where

$$E_r^i = T_r^i \times C_f^i \tag{10.6}$$

where E_r^i denotes the monomial potential ecological risk factor, i.e., for single heavy metals, C_f^i is the contamination factor, T_r^i is the toxic response factor, that indicates the possible hazard and toxicity of specific heavy metals and the environment's vulnerability to pollution. Based on Håkanson's (1980) standardized toxic response factor, T_r^i values for Cd, Cr, Cu, Ni, Pb, and Zn are 30, 2, 5, 5, 5, and 1, respectively.

The classification employed to identify the risk factors E_r^i and R_I was proposed by Håkanson's (1980), wherein:

$E_r^i < 40$ indicates a low potential ecological risk;
$40 < E_r^i < 80$ is a moderate ecological risk;
$80 < E_r^i < 160$ is a considerable ecological risk;
$160 < E_r^i < 320$ is a high ecological risk and.
$E_r^i > 320$ is a very high ecological risk.
$R_I < 95$ indicates a low potential ecological risk;
$95 < R_I < 190$ is a moderate ecological risk;
$190 < R_I < 380$ is a considerable ecological risk and.
$R_I > 380$ is a very high ecological risk.

Finally, the enrichment factor (EF) is determined by dividing the ratio of the metal to the normalizing element in a given sample by the corresponding ratio seen in a reference or baseline environment. Saur and Juste (1994) and Reimann and De Caritat (2000) reported that a reference element (RE) is defined as an element that exhibits limited variability in occurrence and is known for its high stability within the soil. EF levels approaching unity imply a crusted origin, while values below 1.0 signal the potential mobilization of metals. Conversely, EF values greater than 1.0 indicate that the element in concern has anthropogenic origins, whereas a value greater than 10.0 indicates that the elements in concern originate from a non-crusted source (Bhuiyan et al. 2010). Thus, EF is computed using the relationship below,

$$\mathrm{EF} = \frac{(\mathrm{Metal/RE})_{\mathrm{LMFF}}}{(\mathrm{Metal/RE})_{\mathrm{Baseline}}} \tag{10.7}$$

In this study, iron (Fe) was adopted as the RE for geochemical normalization. This selection was made due to the similarity in geochemistry between Fe and several trace elements, as well as the consistent and uniform natural concentration of

Fe, as demonstrated by previous research conducted by Suh and Birch (2005), Tasdemir and Kural (2005), Bhuiyan et al. (2010), and Birch et al. (2020).

10.2.5 Instrumental Characterization

The LMFF was subjected to grinding and homogenization after which the overall composition was determined using an X-ray powder Diffractometer (XRD) (Make: Agilent- single source supernova E). XRD is an efficient analytical method predominantly employed for the identification of phases in crystalline materials, as well as for obtaining data regarding the dimensions of their unit cells. The LMFF was further subjected to Scanning Electron Microscopy (SEM) (Make: Zeiss- Sigma 300). Finally, the CHNS Elemental Analyzer (Make: Elementar vario MACRO cube CHNS Analyzer) was utilized to analyze LMFF, enabling the rapid detection of carbon, hydrogen, nitrogen, and sulfur content in the sample.

10.3 Results and Discussion

10.3.1 Physico-chemical Characterization

The initial characterization of fine fraction for four dumping sites and the value for each parameter are mentioned in Table 10.2. The pH of the fine fraction for all four dumping sites ranged between 5.9 and 7.4 showing the variation from slightly acidic to slightly basic. The % VS and % TOC of the LMFF which have a direct correlation with the organic matter, ranged between 13.9–17.0% and 7.5–9.2%, respectively, for all the four sites. % TOC and % VS have a significant positive correlation and are often used as parameters to assess potential degradation status of landfilled waste (Moody & Townsend 2017). The relatively low % VS values in LMFF from all the sites indicate stable levels signifying most of the degradable organic matter has been consumed (Kayhanian 1995). Moreover, lower values of % TOC and %

Table 10.2 Physico-chemical characterization of LMFF

Parameters	Boragaon	Ghazipur	Bhalswa	Bandhwari
pH	6.7	7.0	5.9	7.4
Volatile solids (%)	17.0	13.9	14.9	15.1
Ash content (%)	83.05	86.1	85.1	84.9
Total organic carbon (%)	9.2	7.5	8.1	8.2
TKN (%)	0.89	1.35	0.72	1.02
Nitrate (%)	1.4	0.6	1.7	0.2
Total nitrogen (%)	2.3	1.9	2.4	1.2
C/N	10.3	5.6	11.3	8.0

VS in the LMFF from Ghazipur demonstrates it has achieved greater degradation of waste in comparison with LMFF from Boragaon. Furthermore, the C/N ratio is also an important factor that indicates the decomposition rate, nitrogen availability, and microbial activity of organic matter in soil or soil conditioner. C/N ratio for LMFF from all four sites ranged 5.6–11.3, and these values indicate that the LMFF from all the sites have achieved maturity and the waste decomposition is advanced and nearing completion. These values also conform with the compost standards, i.e., IS: 16556 (2016), wherein the maximum C/N ratio should not exceed 20 for a compost.

The variation of macro-nutrients present in LMFF from all the dump sites is displayed in Fig. 10.2. LMFF from Bhalswa and Boragaon show relatively high N levels, whereas Boragaon and Ghazipur show relatively higher K levels. Available P content is relatively stable across all the sites, whereas Ca levels, which often originate from inorganic fractions, is highest in Ghazipur and Boragaon. The variation in nutrient dynamics in the LMFF is affected by different factors like age and layering of waste, composition of waste, degradation dynamics of the landfilled waste, and other site-specific factors (Sekhohola-Dlamini and Tekere 2020). However, based on the high nutrient values, it is evident that LMFF from all the dump sites have most likely stabilized or are in their advanced decomposition stage. Based on the physico-chemical characteristics, LMFF can be characterized as a compost-like material rather than soil-like material, which is a term widely reported in literature. The LMFF has the potential to serve as a soil conditioner because of the abundance

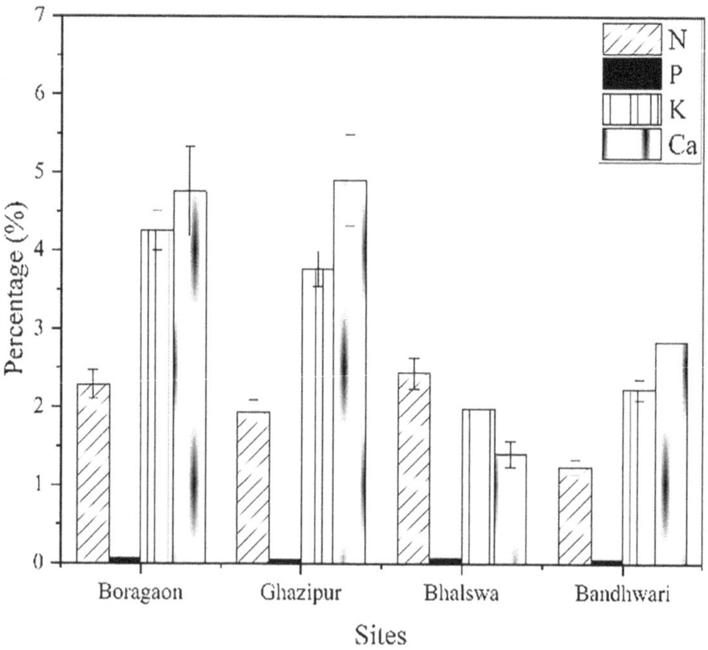

Fig. 10.2 Nutrients in LMFF from different sites

of nutrients and optimal C/N ratio, subject to the quantity of total heavy metals and their subsequent leachability. Moreover, LMFF also has a potential to serve as a viable alternative to local soil, but high amount of organic matter limits its use as a construction material particularly for subgrade and embankment applications.

10.3.2 Total Heavy Metals, and Their Bioavailability and Leachability

Figure 10.3 represents the analysis of total heavy metals (Cr, Co, Cu, Fe, Mn, Ni, Pb, Zn) found in LMFF procured from all four sites along with regulatory standards for application as a compost. The concentration of Co was determined to be the lowest for all the sites, while the concentration of Fe was identified as the highest in LMFF. The total heavy metal concentration found in LMFF from all sites followed

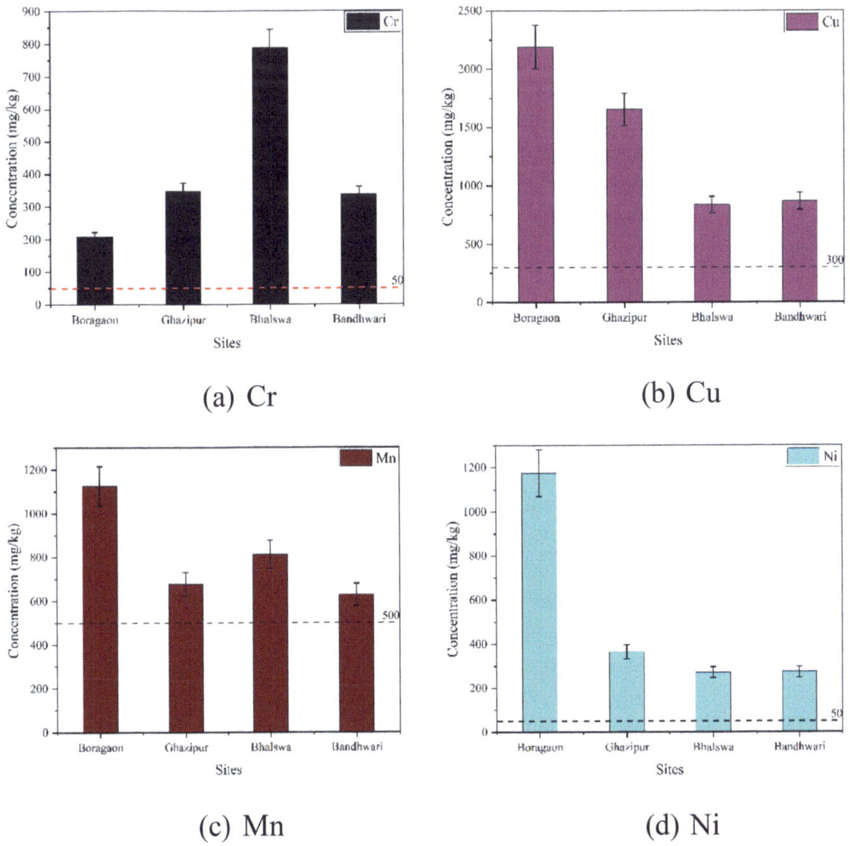

Fig. 10.3 Total heavy metals in LMFF

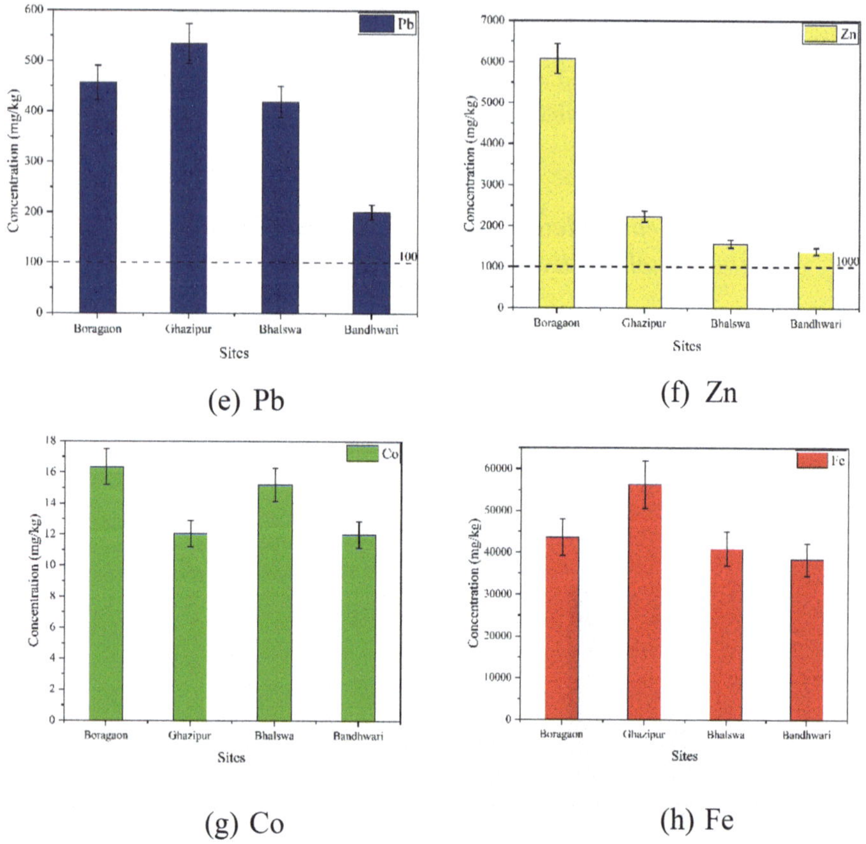

Fig. 10.3 (continued)

the trend: Fe > Zn > Cu. Based on the analysis, it is also evident that all the heavy metals are beyond the permissible limits prescribed by Indian standard for utilization as MSW compost (IS 16556 2016). The majority of the total heavy metal values align with the values documented in the current literature. Prechthai et al. (2008), Quaghebeur et al. (2013), Kaczala Fabio et al. (2015), and Somani et al. (2018) have previously documented elevated levels of Cr in LMFF. The variance in concentration and overall presence of total heavy metals in the LMFF can be attributed to both the composition and age of the waste. It is widely recognized that dumpsites in India predominantly contain a mixture of waste materials that have been accumulated and discarded over an extended period (Narayana 2009).

DTPA, or diethylenetriaminepentaacetic acid, is similar to ethylenediaminetetraacetic acid and is a chelating agent which forms stable water-soluble complexes with metal ions, known as chelates (Kazantzis 2007). This property allows DTPA to enhance the uptake of heavy metals by plants, by increasing their solubility in the DTPA solution. Thus, bioavailability refers to the fraction of heavy metals that are

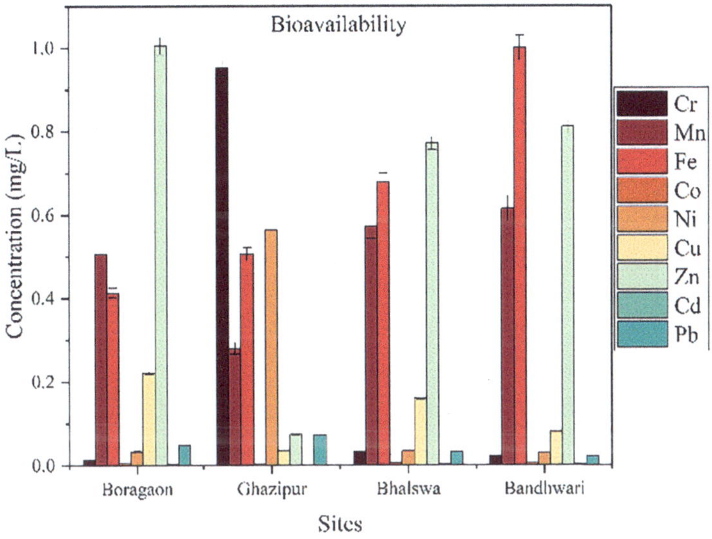

Fig. 10.4 Bioavailability of heavy metals

available for uptake by plants and hence have a possibility of entering food chain, the results of which are depicted in Fig. 10.4. It is evident that, Zn is more bioavailable in comparison with other heavy metals in Boragaon and Bhalswa dumpsites where as Cr and Fe have higher bioavailability in case of Ghazipur and Bandhwari dumpsites respectively. The higher bioavailability of toxic metals especially Cr in Ghazipur raises concerns pertaining to human health and ecosystems. Moreover, it is also evident that bioavailability of different heavy metals in different sites reflect the differences in landfill conditions such as pH, redox potential, waste composition, and degradation dynamics as observed in study by Sun et al. (2024).

Leachability refers to the tendency of heavy metals to migrate from their source material and potentially pollute the surrounding environment. In instances where toxic heavy metals are leached into the surrounding area and exceed the regulatory threshold values, these heavy metals may be classified as environmentally hazardous and present a probable threat to human health. Based on the leachability extraction (Fig. 10.5), the Fe was the highest leachable heavy metal from Boragaon, Bhalswa, and Bandhwari sites and Cu was the highest leachable heavy metal from Ghazipur. However, based on the EPA (1992) regulatory levels, the heavy metals in the leachate from all the dump sites did not exceed the threshold values, and hence LMFF and its leachate from all the sites are classified as non-hazardous in nature. However, these leachable heavy metals still have the potential to contaminate the groundwater and geoenvironment. pH, organic matter, and other site-specific conditions affect the leachability of heavy metals and hence, it is important that LMFF from all the dumpsites be stabilized using effective mitigation strategies.

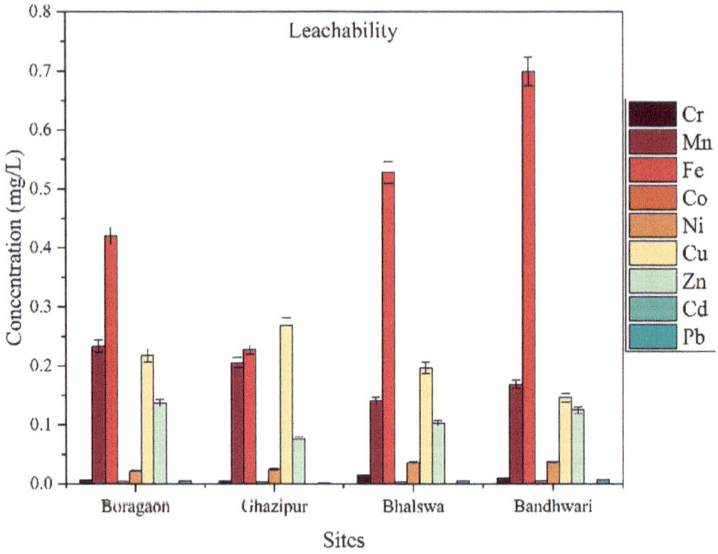

Fig. 10.5 Leachability of heavy metals

10.3.3 Pollution Indices

Table 10.3 displays the contamination factor and pollution load index for heavy metals in LMFF obtained from the four sites. It is evident that Cu, Pb, and Zn in LMFF cause high contamination across all the dump sites whereas Cr in Bhalswa and Ni in Boragaon also cause high contamination with $C_f^i \geq 6$. Based on the PLI, Boragaon is the most contaminated site driven by Cu, Ni, Pb, and Zn concentrations followed by Ghazipur, Bhalswa, and Bandhwari sites. The elevated C_f^i and PLI values in the LMFF from all the dumpsites underscore the need for treatment of LMFF to mitigate the long-term impact of heavy metal contamination of soil and groundwater systems.

Similarly, Table 10.4 presents the potential ecological risk assessment for LMFF from the four sites. The monomial ecological risk factors (E_r^i) display a high ecological risk from Cu in Boragaon and Ghazipur whereas a considerable ecological risk in Bhalswa and Bandhwari. Similarly, Pb also causes a considerable ecological risk in Boragaon, Ghazipur, and Bhalswa sites. Furthermore, based on the potential ecological risk index (R_I)- which is a cumulative measure of all the monomial factors- puts LMFF from Boragaon and Ghazipur under a very high-risk category and Bhalswa and Bandhwari under a considerable risk category. These differences could be attributed to variables such as waste composition, age of dumpsite, waste degradation, and other geographic and climatic conditions. Irrespective of these differences, heavy metals like Cu, Pb, and Ni are the major contributors to the risk index across all the sites necessitating stabilization measures for LMFF to mitigate the long-term impacts.

Table 10.3 Contamination factor (C_f^i) and PLI

Heavy metals	Contamination factor, C_f^i			
	Boragaon	Ghazipur	Bhalswa	Bandhwari
Cr	3.43	5.75	13.03	5.58
Fe	1.23	1.59	1.15	1.08
Cu	48.21	36.45	18.38	19.04
Ni	20.13	6.24	4.66	4.69
Pb	23.41	27.41	21.49	10.32
Zn	68.28	25.03	17.54	15.57
Mn	1.72	1.03	1.24	0.96
Pollution load index	**10.17**	**7.61**	**6.69**	**5.04**

Table 10.4 Potential ecological risk assessment

Heavy metals	Monomial potential ecological risk factor, E_r^i			
	Boragaon	Ghazipur	Bhalswa	Bandhwari
Cr	6.86	11.49	26.05	11.15
Cu	241.04	182.25	91.90	95.22
Ni	120.78	37.46	27.93	28.13
Pb	117.06	137.05	107.45	51.58
Zn	68.28	25.03	17.54	15.57
Potential ecological risk index, R_I	**554.03**	**393.29**	**270.87**	**201.65**
Risk category	**Very high risk**	**Very high risk**	**Considerable risk**	**Considerable risk**

Table 10.5 Enrichment factor for LMFF

Heavy metals	Enrichment factor, EF			
	Boragaon	Ghazipur	Bhalswa	Bandhwari
Cr	2.79	3.62	11.28	5.15
Fe	1.00	1.00	1.00	1.00
Cu	39.16	22.97	15.92	17.59
Ni	16.35	3.93	4.03	4.33
Pb	19.02	17.27	18.61	9.53
Zn	55.46	15.77	15.19	14.39
Mn	1.40	0.65	1.07	0.89

Table 10.5 presents the enrichment factor for LMFF across all the dumpsites. Based on this, LMFF from Boragaon shows high EF (>10) values for Zn, Cu, Pb, and Ni indicating substantial anthropogenic inputs. LMFF from Ghazipur and Bandhwari also exhibit significant enrichment for Cu, Zn, and Pb, whereas Bhalswa demonstrated higher enrichment for Pb, Cu, Zn, and Cr. Higher EF values for LMFF across all the dumpsites highlight the potential of LMFF to cause soil contamination and pose risks for the surrounding ecosystems.

10.3.4 Instrumental Characterization

The identification of compounds in the LMFF recovered from landfill mining across four dumpsites were performed using the XRD and the respective peaks observed are depicted in Fig. 10.6. The graph illustrates a significant presence of silica (SiO_2) peaks or quartz within the LMFF across all the dumpsites. The elevated levels of silica and alumina compounds seen in the LMFF may be attributed to the presence of construction and demolition waste, particulate matter, and fragmented glass residues within the legacy waste. Comparable results were obtained in studies conducted by Jahagirdar et al. (2026) and Rawat and Mohanty (2021) wherein the samples consisted of major inorganic peaks of quartz, calcite, muscovite, and gypsum. Thus, XRD analysis highlights the fact that LMFF across all the sites also contain significant inorganics along with organic matter. Furthermore, Fig. 10.7 portrays the morphology of the LMFF across the four sites using SEM which illustrate the irregular and non-uniform nature of the particles in LMFF. Moreover, the presence of needle-like and thread-like particles observed most likely depicts the presence of plastic/fibers in LMFF. This image-processing method confirms the higher organic porosity and heterogeneity of the LMFF across all the dumpsites.

Finally, Table 10.6 depicts the results from CHNS analysis. Lower C/N ratio, as observed across all the sites, suggests relatively advanced decomposition due to

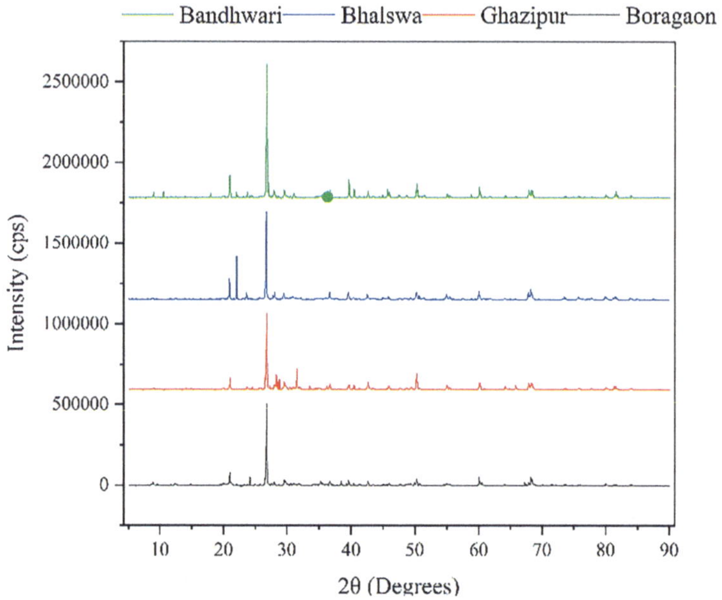

Fig. 10.6 XRD analysis of LMFF

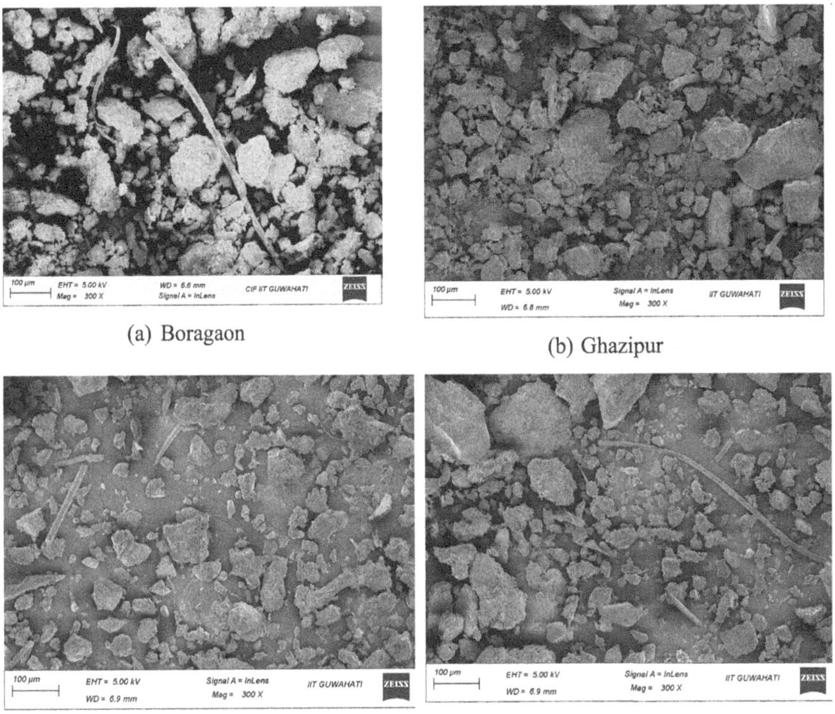

(a) Boragaon (b) Ghazipur (c) Bhalswa (d) Bandhwari

Fig. 10.7 SEM images of LMFF

Table 10.6 CHNS analysis of LMFF

Parameters	Boragaon	Ghazipur	Bhalswa	Bandhwari
N (%)	1.02	0.44	0.50	0.89
C (%)	12.35	6.11	7.04	10.33
H (%)	1.87	1.07	1.02	1.22
S (%)	1.78	1.86	1.28	1.02
C/N ratio	12.11	14.01	14.10	11.61
C/H ratio	6.62	5.72	6.91	8.47

nitrogen retention. A higher C/H ratio also suggests the presence of more recalcitrant, aromatic carbon structures, often associated with older or stabilized organic matter. These values indicate that the LMFF from all the sites have achieved maturity and the waste decomposition is advanced and nearing completion. The physicochemical characterization of LMFF also yields similar results, confirming the presence of a significant amount of organic matter and nutrient content.

10.4 Potential Valorization of LMFF and Future Prospects

The present study reveals the distinct characteristics of LMFF. The physico-chemical characterization reveals that the LMFF across all the four dumpsites have exceptional nutritional composition with the majority of parameters conforming to compost specifications (National Centre of Organic Farming 2014). Moreover, the results also demonstrate that LMFF from all the sites have achieved maturity and the waste decomposition is advanced and nearing completion. The results also demonstrate the presence of inorganics along with organic matter in LMFF from all the sites. However, the presence of elevated levels of total heavy metals limits its potential use as a soil conditioner or construction material in civil engineering applications (Singh and Chandel 2023). Considering the evident ability of the LMFF to cause heavy metal contamination in soil and surrounding ecosystem, it is crucial to emphasize the treatment and valorization of LMFF rather than its mere disposal at the dump sites. The treatment methods might include,

- Immobilization of heavy metals in LMFF by the application of lime, biochar, or fly ash. Various studies like Gray et al. (2006), Malik et al. (2018), and Haghighatjou & Shirvani (2020) have shown the treatment of metal-contaminated soils or acidic soils employing the application of lime or biochar to raise the pH. The addition of lime or biochar could also increase its ability to retain heavy metals because of a high surface area.
- Treatment with phytoextraction or phytoremediation techniques which employ plants—known as hyperaccumulators to extract heavy metals from soil or waste (Brown et al. 1994). These plants either absorb the heavy metals in their tissues or promote chemical reactions to immobilize them (Yang et al. 2022), and the potential for this method has been reported by Jahagirdar et al. (2026)
- Heavy metal stabilization of LMFF and reuse by means of geopolymerization- a process known to immobilize the heavy metals by binding them into the geopolymer matrix (Van Jaarsveld et al. 1998). The geopolymerized products could be utilized in civil engineering applications thereby serving the purpose of circular economy.
- Microbial and enzymatic treatment of the LMFF to encapsulate the heavy metals by microbe-metal interactions which entail biosorption, bioaccumulation, or biomineralization, all of which result in stable complexes that effectively sequester heavy metals (Ayangbenro and Babalola 2017).

Future studies could explore the efficiency of these treatment methods on LMFF. Based on the leaching characteristics and immobilization of heavy metals from LMFF after treatment, it could be used in off-site applications such as a soil conditioner for inedible crops, as a road-base or backfill material, as landfill cover, or in other civil engineering applications. The future studies could also focus on evaluating the economic feasibility of large-scale implementation of treating LMFF. This could include cost-benefit analyses on the different treatment methods, considering factors such as operational expenses, scalability, and market demand for valorized LMFF products. LMFF's revenue generation through circular economy framework could also be explored wherein LMFF-based products are promoted thereby reducing dependency on primary resources.

10.5 Conclusions

In conclusion, this study provides valuable insights into the distinctive characteristics of fine fraction from all the four sites and its potential treatment methods. In the broader context, this study highlights several key findings, including,

- The LMFF from all sites have high organic matter (13–17%) with the NPK values of the LMFF similar to those found in city compost, indicating a substantial presence of macro-nutrients.
- LMFF from all sites also exhibited elevated levels of total heavy metal concentrations exceeding regulatory limits for utilization as MSW compost, with Cr, Zn, and Fe having greater bioavailability and potential to readily enter the food chain.
- The potential ecological risk index (R_I) puts LMFF from Boragaon and Ghazipur under a very high-risk category and Bhalswa and Bandhwari under a considerable risk category. The differences in risk indices could be attributed to variables such as waste composition, age of dumpsite, waste degradation dynamics, and other geographic and climatic conditions.

Given the evident environmental implications of LMFF, it is imperative to advocate for its treatment followed by utilization rather than its mere disposal in low-lying areas. The ecological risk posed by LMFF emphasizes the urgency of implementing appropriate treatment methods. Raising the pH levels by application of lime/biochar, phytoremediation techniques, geopolymerization, and microbial/enzymatic treatments are some of the proposed strategies to immobilize heavy metals and enhance LMFF's environmental compatibility. These actions shall help promote public health and environmental integrity by reducing the risk of contamination from heavy metals and other unexplored contaminants. Treatment followed by utilization of LMFF can lead to economic benefits through resource recovery and waste-to-wealth initiatives. To comprehensively understand the nature of LMFF, future studies could also delve into presence and treatment of contaminants of emerging concern, ensuring a holistic approach to its utilization.

Acknowledgments The authors are grateful to the Department of Civil Engineering and the Central Instrumentation Facility (CIF) at the Indian Institute of Technology Guwahati, India for providing workspace and conducting instrumental characterization (XRD, SEM, CHNS, and ICPMS).

Funding The authors would like to acknowledge the funding received (Project No- DST/TDT/TDP-30/2022) from the Department of Science and Technology, Government of India under the Technology Development Transfer Division and Technology Development Programme.

References

American Public Health Association (1992) APHA method 4500-NO3: standard methods for the examination of water and wastewater

Ayangbenro AS, Babalola OO (2017) A new strategy for heavy metal polluted environments: a review of microbial biosorbents. Int J Environ Res Public Health 14(1) MDPI. https://doi.org/10.3390/ijerph14010094

Bhuiyan MAH, Parvez L, Islam MA, Dampare SB, Suzuki S (2010) Heavy metal pollution of coal mine-affected agricultural soils in the northern part of Bangladesh. J Hazard Mater 173(1–3):384–392. https://doi.org/10.1016/j.jhazmat.2009.08.085

Birch GF, Lee JH, Tanner E, Fortune J, Munksgaard N, Whitehead J, Coughanowr C, Agius J, Chrispijn J, Taylor U, Wells F, Bellas J, Besada V, Viñas L, Soares-Gomes A, Cordeiro RC, Machado W, Santelli RE, Vaughan M et al (2020) Sediment metal enrichment and ecological risk assessment of ten ports and estuaries in the world Harbours project. Mar Pollut Bull 155. https://doi.org/10.1016/j.marpolbul.2020.111129

Brown SL, Chaney RL, Angle JS, Baker AJM (1994) Phytoremediation potential of Thlaspi caerulescens and bladder campion for zinc- and cadmium-contaminated soil. J Environ Qual 23(6):1151–1157. https://doi.org/10.2134/jeq1994.00472425002300060004x

Burlakovs J, Pilecka J, Grinfelde I, Arbidans L, Arina D, Setyobudi RH (2021) Sustainable landfill fine fraction of waste reuse opportunities in covering layer development. Res Rural Develop 36:303–310. https://doi.org/10.22616/rrd.27.2021.043

EPA, E. P (1992) Method 1311: toxicity characteristic leaching procedure. US Environmental Protection Agency

Gray CW, Dunham SJ, Dennis PG, Zhao FJ, McGrath SP (2006) Field evaluation of in situ remediation of a heavy metal contaminated soil using lime and red-mud. Environ Pollut 142(3):530–539. https://doi.org/10.1016/j.envpol.2005.10.017

Haghighatjou M, Shirvani M (2020) Sugarcane bagasse biochar: preparation, characterization, and its effects on soil properties and zinc sorption-desorption. Commun Soil Sci Plant Anal 51(10):1391–1405. https://doi.org/10.1080/00103624.2020.1763383

Hakanson L (1980) An ecological risk index for aquatic pollution control. A sedimentological approach. Water Res 14(8):975–1001. https://doi.org/10.1016/0043-1354(80)90143-8

Hogland W (2002) Remediation of an old landsfill site soil analysis, leachate quality and gas production William Hoglan d. ESPR-Environ Sci Pollut Res 1:49–54. http://dx.doi.ocr

IS 16556 (2016) Municipal solid waste compost, manure grade-specification. www.standardsbis.in

Jahagirdar, V., Maturi, K. C., Mishra, A. K., & Kalamdhad, A. S. (2025). Characterization and pollution risk assessment of landfill-mined fine fraction: a case study from Guwahati, India. *Environmental Monitoring and Assessment*, 197(7), 1–18.

Jahagirdar, V., Mishra, A. K., & Kalamdhad, A. S. (2026). From Waste to Harvest: A Case Study on the Growth Dynamics of Vigna radiata L. and Health Implications of Solanum lycopersicum L. Cultivated in Landfill-Derived Soil. *Journal of Hazardous, Toxic, and Radioactive Waste*, 30(1), 04025046.

Jain P, Kim H, Townsend TG (2005) Heavy metal content in soil reclaimed from a municipal solid waste landfill. Waste Manag 25(1):25–35. https://doi.org/10.1016/j.wasman.2004.08.009

Jones PT, Geysen D, Tielemans Y, Van Passel S, Pontikes Y, Blanpain B, Quaghebeur M, Hoekstra N (2013) Enhanced landfill mining in view of multiple resource recovery: a critical review. J Clean Prod 55:45–55. https://doi.org/10.1016/j.jclepro.2012.05.021

Joseph, K., Visvanathan, C., & Hogland, W. (2008). Dumpsite rehabilitation manual
Kaartinen T, Sormunen K, Rintala J (2013) Case study on sampling, processing and characterization of landfilled municipal solid waste in the view of landfill mining. J Clean Prod 55:56–66. https://doi.org/10.1016/j.jclepro.2013.02.036
Fabio K, Hadi MM, Allar L (2015) Leaching characteristics of the fine fraction from an excavated landfill: physico-chemical characterization. J Mater Cycles Waste Manag 19:294–304
Kaushal RK, Chabukdhara M, Varghese GK (2012) Municipal solid waste management in India-Current State and Future Challenges: a review. Int J Eng Sci Technol. https://www.researchgate.net/publication/233894305
Kayhanian M (1995) Biodegradability of the organic fraction of municipal solid waste in a high-solids anaerobic digester. Waste Manag Res 13(2):123–136. https://doi.org/10.1016/S0734-242X(95)90114-0
Kazantzis G (2007) Diagnosis and treatment of metal poisoning—general aspects. In: Handbook on the toxicology of metals, 3rd edn, pp 303–317. https://doi.org/10.1016/B978-012369413-3/50070-7
Kennedy N, Lally RD, Walsh SW, Dowling DN, Ryan D (2021) Effect of green waste and lime amendments on biostabilisation, physical-chemical and microbial properties of the composted fine fraction of residual municipal solid waste. Waste Manag Res 39(8):1069–1077. https://doi.org/10.1177/0734242X21996823
Maanan M, Saddik M, Maanan M, Chaibi M, Assobhei O, Zourarah B (2015) Environmental and ecological risk assessment of heavy metals in sediments of Nador lagoon, Morocco. Ecol Indic 48:616–626. https://doi.org/10.1016/j.ecolind.2014.09.034
Malik Z, Yutong Z, ShengGao L, Abassi GH, Ali S, Imran khan M, Kamran M, Jamil M, Al-Wabel MI, Rizwan M (2018) Effect of biochar and quicklime on growth of wheat and physicochemical properties of Ultisols. Arab J Geosci 11(17). https://doi.org/10.1007/s12517-018-3863-1
Moody CM, Townsend TG (2017) A comparison of landfill leachates based on waste composition. Waste Manag 63:267–274. https://doi.org/10.1016/j.wasman.2016.09.020
Narayana T (2009) Municipal solid waste management in India: from waste disposal to recovery of resources? Waste Manag 29(3):1163–1166. https://doi.org/10.1016/j.wasman.2008.06.038
National Centre of Organic Farming (2014) Biofertilizers and organic fertilizers in fertilizer control order
Prechthai T, Padmasri M, Visvanathan C (2008) Quality assessment of mined MSW from an open dumpsite for recycling potential. Resour Conserv Recycl 53(1–2):70–78. https://doi.org/10.1016/j.resconrec.2008.09.002
Quaghebeur M, Laenen B, Geysen D, Nielsen P, Pontikes Y, Van Gerven T, Spooren J (2013) Characterization of landfilled materials: screening of the enhanced landfill mining potential. J Clean Prod 55:72–83. https://doi.org/10.1016/j.jclepro.2012.06.012
Rawat P, Mohanty S (2021) Experimental investigation on MSW fine mixed with fibers: fiber reinforced waste. J Hazard Toxic Radioact Waste 25(3):04021009. https://doi.org/10.1061/(asce)hz.2153-5515.0000609
Reddy CS, Mohanty S, Shaik R (2018) Physical, chemical and geotechnical characterization of fly ash, bottom ash and municipal solid waste from Telangana state in India. Int J Geo-Eng 9(1). https://doi.org/10.1186/s40703-018-0093-z
Reimann C, De Caritat P (2000) Intrinsic flaws of element enrichment factors (EFs) in environmental geochemistry. Environ Sci Technol 34(24):5084–5091. https://doi.org/10.1021/es001339o
Saur E, Juste C (1994) Enrichment of trace elements from long-range aerosol transport in Sandy Podzolic soils of Southwest France. Water Air Soil Pollut 73:235–246
Sekhohola-Dlamini L, Tekere M (2020) Microbiology of municipal solid waste landfills: a review of microbial dynamics and ecological influences in waste bioprocessing. Biodegradation 31(1–2) Springer. https://doi.org/10.1007/s10532-019-09890-x
Singh A, Chandel MK (2020) Effect of ageing on waste characteristics excavated from an Indian dumpsite and its potential valorisation. Process Saf Environ Prot 134:24–35. https://doi.org/10.1016/j.psep.2019.11.025

Singh A, Chandel MK (2022) Valorization of fine fraction from legacy waste as fired bricks: a step towards circular economy. J Clean Prod 331:129918. https://doi.org/10.1016/j.jclepro.2021.129918
Singh A, Chandel MK (2023) Physicochemical and biological assessment of legacy waste for application as soil conditioner. Environ Sci Pollut Res 30(11):29699–29710. https://doi.org/10.1007/s11356-022-24295-9
Singh, J., & Kalamdhad, A. (2019). Bioavailability, leachability, chemical speciation, and bioremediation of heavy metals in the process of composting
Singh J, Kalamdhad AS (2012) Concentration and speciation of heavy metals during water hyacinth composting. Bioresour Technol 124:169–179. https://doi.org/10.1016/j.biortech.2012.08.043
Singh WR, Kalamdhad AS (2016) Transformation of nutrients and heavy metals during vermicomposting of the invasive green weed Salvinia natans using Eisenia fetida. Int J Recycl Org Waste Agric 5(3):205–220. https://doi.org/10.1007/s40093-016-0129-3
Somani M, Datta M, Ramana GV, Sreekrishnan TR (2018) Investigations on fine fraction of aged municipal solid waste recovered through landfill mining: case study of three dumpsites from India. Waste Manag Res 36(8):744–755. https://doi.org/10.1177/0734242X18782393
Suh J-Y, Birch GF (2005) Use of grain-size and elemental normalization in the interpretation of trace metal concentrations in soils of the reclaimed area adjoining Port Jackson, Sydney, Australia. Water Air Soil Pollut 160;357–371
Sun, Q., Sun, B., Wang, D., Pu, Y., Zhan, M., Xu, X., Wang, J., & Jiao, W. (2024). A review on the chemical speciation and influencing factors of heavy metals in municipal solid waste landfill humus. Waste Dispos Sustain Energy, 1–10
Tasdemir Y, Kural C (2005) Atmospheric dry deposition fluxes of trace elements measured in Bursa, Turkey. Environ Pollut 138(3):462–472. https://doi.org/10.1016/j.envpol.2005.04.012
Tomlinson DL, Wilson JG, Harris CR, Jeffrey DW (1980) Problems in the assessment of heavy-metal levels in estuaries and the formation of a pollution index. Helgol Meeresunters 33:566
Van Jaarsveld JGS, Van Deventer JSJ, Lorenzen L (1998) Factors affecting the immobilization of metals in Geopolymerized Flyash. Metall Mater Trans B 29:283
Wanka S, Münnich K, Fricke K (2017) Landfill mining—wet mechanical treatment of fine MSW with a wet jigger. Waste Manag 59:316–323. https://doi.org/10.1016/j.wasman.2016.10.050
Yang Z, Yang F, Liu JL, Wu HT, Yang H, Shi Y, Liu J, Zhang YF, Luo YR, Chen KM (2022) Heavy metal transporters: functional mechanisms, regulation, and application in phytoremediation. Sci Total Environ 809). Elsevier B.V. https://doi.org/10.1016/j.scitotenv.2021.151099
Zhou C, Gong Z, Hu J, Cao A, Liang H (2015) A cost-benefit analysis of landfill mining and material recycling in China. Waste Manag 35:191–198. https://doi.org/10.1016/j.wasman.2014.09.029

Chapter 11
Field Investigation on Repeated Trommeling for Improved Quality of Segregated Legacy Waste: Suitability for Bulk Off-Site Reuse

Debaprakash Parida, G. V. Ramana, and Manoj Datta

11.1 Introduction

The global trajectory of municipal solid waste (MSW) generation is alarming, with an estimated 2 billion tons produced annually, projected to increase by 70% by 2050 (Lama 2024). Inadequate waste collection infrastructure, coupled with the widespread reliance on unmanaged landfills and open dumps for the disposal of mixed MSW, exacerbates the challenges in effective waste management. These aging, toxic waste mountains, often referred to as legacy waste, have become entrenched in urban landscapes, significantly contributing to global crises, including climate change, pollution, and biodiversity loss. The harmful emissions emanating from these open dumps, which infiltrate both terrestrial and aquatic ecosystems, are linked to a multitude of adverse environmental and public health consequences, some of which may persist for generations (Siddiqua et al. 2022; United Nations Environment Programme 2024; Vinti et al. 2021; Wilson et al. 2022). If left unaddressed, the ongoing emissions from uncontrolled waste disposal systems will continue to undermine environmental integrity, making the achievement of the United Nations Sustainable Development Goals (UNSDG) increasingly untenable (Lama 2024).

Recognizing waste and resource management as pivotal to sustainable development and climate change mitigation is critical. The current challenge posed by unmanaged legacy waste, though dire, is not insurmountable. Existing strategies to mitigate the planetary crisis caused by legacy waste prioritize resource recovery and embrace circular economy principles, focusing on recycling and reuse to minimize environmental impact (Lama 2024). Addressing global climate change, pollution, and biodiversity loss necessitates a global and local dialog on waste management

D. Parida (✉) · G. V. Ramana · M. Datta
Department of Civil Engineering, Indian Institute of Technology Delhi, New Delhi, India

M. Somani et al. (eds.), *Landfill Mining of Legacy MSW: Global Insights and Indian Experience*, Springer Transactions in Civil and Environmental Engineering, https://doi.org/10.1007/978-981-95-3959-8_11

and more proactive actions. In the context of circular economy and resource recovery, landfill mining (LFM) has emerged as a sustainable approach, involving the excavation of legacy waste to recover recyclable and reusable materials while reducing landfill footprint (Datta et al. 2021; Hölzle 2018; Krook and Baas 2013; Kurian et al. 2003). Enhanced landfill mining (ELFM) builds upon this concept by incorporating waste-to-materials (WtM) and waste-to-energy (WtE) strategies, aiming to valorize waste into high-value products and energy (De Vocht and Descamps 2014; Esguerra et al. 2020; Frändegård et al. 2015; Hogland and Marques 2010; Joakim Krook 2021; Zhang et al. 2019).

ELFM is increasingly adopted globally as a strategy to mitigate the environmental impacts of unmanaged waste dumps, reclaim land occupied by waste, and recover valuable resources for repurposing and reintegration into the supply chain (Abdel-Shafy and Mansour 2018; Danthurebandara 2016; Jones et al. 2013; Parrodi et al. 2019; Quaghebeur et al. 2013). With approximately 40% of global MSW disposed of in open dumps, ELFM offers significant potential for resource recovery and environmental remediation (United Nations Environment Programme 2024; Wilson et al. 2022).

According to the Swachh Bharat Mission 2.0 (SBM 2.0) dashboard, over 2300 lakh tons of legacy waste are stored in more than 2400 open dumps across India (Ministry of Housing and Urban Affairs 2025). Since, the introduction of the Waste Management Rules in 2016, LFM has gained traction in India as a method to recover valuable resources from legacy dumps, convert waste to engineered landfills, and reclaim land occupied by waste (CPCB 2019). Figure 11.1 illustrates the ongoing LFM operation at the studied dumpsite.

In India, LFM currently involves manual segregation of oversized fractions followed by mechanical segregation, commonly known as trommeling, using 30

Fig. 11.1 Ongoing LFM operation at the studied dumpsite

and 6 mm trommels. These trommels are named after the perforation sizes of the screens attached to them. The trommeling process separates legacy waste into +30 mm fractions (including boulders and refuse-derived fuels, RDF) and −30 mm fractions, which are further segregated through a 6 mm trommel into gravel-like material (GLM) of 30 to 6 mm (coarse fraction, CF) and soil-like material (SLM) of −6 mm (fine fraction, FF). However, most LFM projects in India adopt a rapid approach that does not align with prescribed guidelines (CPCB 2019) driven by space constraints and pressure for quick waste removal. This results in low-quality segregated legacy waste (SLW), with high contamination levels and limited reuse potential. Despite significant CF and FF generation, their use in off-site applications is hindered by quality concerns and inadequate disposal land. Therefore, evaluating the quality of SLW before reuse is crucial.

Extensive research has focused on assessing the contamination potential of FF and treatment methods to enhance their quality for reuse as a substitute for natural soils in off-site applications. However, considering the large volume of legacy waste being processed, a substantial amount of CF in the 30–6 mm range (>20% of total processed legacy waste) can be generated through LFM, potential for repurposing as construction materials. This study, the first of its kind in India, investigates a simple yet innovative technique to improve the quality of contaminated CF segregated from legacy waste through a full-scale field approach. The hypothesis is that fine fractions have a high contamination potential (Parida et al. 2024) and removing these fines from the CF can enhance its quality. By investigating the reuse potential of segregated CF, this study aims to recover secondary resources that reduce the dependency on natural ones, contributing to environmental sustainability and supporting the circular economy. The research provides insights into India's current approach to segregating legacy waste through rapid trommeling, its impact on the quality of segregated fractions, and the potential of field-scale approaches to improve legacy waste quality, focusing on the recovery of high-quality secondary resources from MSW dumps. The specific objectives of this study are: (a) **Characterizing SLW**—to understand its physical and chemical properties, contamination potential, and variability, essential for optimizing the segregation process and improving resource recovery efficiency; (b) **Evaluating the effectiveness of the field-scale approach**—to develop a feasible and innovative methodology for treating contaminated CF and identifying sustainable methods for resource recovery; (c) **Assessing CF suitability for bulk reuse**—to evaluate the quality of CF generated from repeated dry trommeling, determining the potential of the field-scale approach to improve the quality of segregated legacy waste.

In the absence of clear guidelines for SLW reuse, this comprehensive assessment can assist policymakers and governments in developing appropriate frameworks and guidelines for its off-site applications.

11.2 Material and Methods

11.2.1 Site Description

Since 2019, landfill mining has been operational at three major open dumpsites in Delhi, India, following directives from the National Green Tribunal (NGT). The dumpsite considered for the present study, which is the largest among the three, ranks among the 50 largest dumpsites globally (Waste Atlas 2014). Figure 11.2 illustrates the dumpsite location and its surrounding environment.

For the past four decades, mixed MSW has been deposited at this site, which now holds an estimated 140 lakh tons of legacy waste, reaching a height of 65 meters above ground level as of 2019. The dumpsite spans 70 acres, surrounded by densely populated human settlements, with an open channel drain approximately 50 meters wide running adjacent to the site. Currently, the dumpsite receives 1776 tons of fresh MSW daily, further expanding its size. Landfill mining operations have been ongoing at the site since July 2019, where legacy waste is excavated and immediately segregated through trommeling at the dumpsite. As of September 2024, a total of 27 lakh tons of legacy waste have been processed, and the site is projected to be fully remediated by December 2028 (Municipal Corporation of Delhi 2024).

11.2.2 Sampling and Processing

In this study, the field approach to improve the quality of segregated CF was preceded by a thorough physical characterization and assessment of contamination potential and contaminants of concern (CoC) in the SLW fractions obtained from

Fig. 11.2 Location of the dumpsite and the sampling points. (Source: Google Earth)

the site. Freshly excavated mixed legacy waste (MLW) and SLW fractions, including the entire trommeled fraction (ETF, −30 mm), CF, and FF samples, were collected from four distinct trommeling locations during the LFM operation at the site. More than 2 tons of MLW and 1 ton each of ETF and CF samples were obtained from each of the four sampling points (refer to Fig. 11.2 for the location of sampling points), totalling 8 tons of MLW and 4 tons of ETF and CF for physical analysis. For chemical analysis, ETF, CF, and FF samples of appropriate quantities were sealed in airtight polyethylene packets and promptly transported to the laboratory, where they were stored in an incubator at 4 °C to preserve their structural integrity. For physical analysis, the freshly excavated moist MLW samples were spread in thin layers and air-dried for 5–6 days until they were suitable for further examination. Representative samples were obtained using the coning and quartering method. Local soil samples, characterized as Delhi silt unaffected by anthropogenic influences, were collected from 0.5 to 1 m depth, up to 5 km from the dumpsite's periphery. These local soil parameters served as reference background levels for comparative contaminant assessment in the SLW fractions.

11.2.3 Characterization of Legacy Waste

Given that the wastes have been deposited at the site for the past four decades without detailed records of their constitutional composition or quantities disposed of by the municipalities and considering that the mixed legacy waste is now being segregated into various components and size fractions through trommeling, it is crucial to characterize these mixed legacy MSWs. This characterization is essential for making informed decisions regarding their potential for recycling and repurposing in sustainable applications.

The composition of the MLW provides detailed insights into the content of individual components that can potentially be recycled and recovered. For compositional analysis, a representative sample of over 200 kg was obtained through coning and quartering of dried MLW, as outlined in ASTM D5231 2016. This sample was then screened using a 20 mm rectangular mesh to segregate the finer fractions that could not be manually sorted into individual components. Components larger than 20 mm were manually sorted into distinct categories, including construction and demolition (C&D) debris, plastics, glass, metals, ceramics, textiles, wood, and others. The percentage content of each component was then calculated and reported.

The grain size distribution (GSD) of the MLW and SLW samples was determined following ASTM D6913 2017. A series of coarse and fine sieves were employed to quantify the percentage content of various size fractions. Washed sieving was conducted to ensure accurate accounting of the fine fractions that could have otherwise adhered to the surface of the coarse fractions during dry sieve analysis.

The moisture content (MC) of the SLW fractions was determined using the oven-drying method, as outlined by Hull et al. 2005; Jani et al. 2016. The freshly collected samples were immediately subjected to MC determination by placing them in a thermostatically controlled oven at 105 °C for 24 h. MC was calculated using the gravimetric method by measuring the mass difference between the moist and oven-dried samples.

The organic content (OC) of the SLW fractions was determined using the loss on ignition (LOI) method, as outlined in ASTM D2974 2014. In this method, SLW samples were heated at 550 ± 50 °C in a muffle furnace for 3 ± 0.5 h. The OC was calculated based on the mass loss resulting from the ignition process. A similar approach was adopted by Burlakovs et al. (2016) and Somani et al. (2020) for determining the OC of legacy waste using LOI, due to its simplicity and reliability.

The leachability of contaminants from the SLW fractions was assessed using a single-stage batch leaching (SSBL) procedure, following the Swedish standards SS-EN 12457-2, (2002) and SS-EN 12457-4 (2002), with deionized (DI) water as the leachant. In this procedure, 100 grams of oven-dried SLW sample were mixed with DI water at a liquid-to-solid (L/S) ratio of 10 L/kg in a 2-L glass bottle. To ensure a homogeneous mixture, the bottle was agitated using an end-to-end rotary agitator at a rotation speed of 30 rpm for 16 ± 2 h. The resulting water extracts, referred to as leachates, were subjected to a three-stage filtration process: centrifugation at 10,000 rpm for 15 min, vacuum filtration through a 2.5 μm Whatman 42 filter paper, followed by filtration through a 0.2 μm cellulose filter to obtain a clear eluate free of suspended particles. The filtered eluates were then analyzed for pH, electrical conductivity (EC), color, total dissolved solids (TDS), water-soluble sulphates and chlorides, and key cations (sodium, magnesium, potassium, calcium). Additionally, trace metals including arsenic (As), chromium (Cr), copper (Cu), nickel (Ni), lead (Pb), and zinc (Zn) were quantified.

Total dissolved solids were determined using the gravimetric method, where 50 mL of leachate sample was dried in an oven at 105 °C for 24 h, and the mass of residual salts was quantified. pH and electrical conductivity (EC) of the leachate samples were measured using high-accuracy electronic probes. The color intensity of the leachate was determined in platinum-cobalt units (PCU) using a tintometer. Water-soluble sulphate concentration was determined by the turbidimetric method, while chloride levels were quantified through argentometric titration with silver nitrate solution, following the procedures outlined in APHA/AWWA/WEF (2012).

Leachable trace metals were analyzed using inductively coupled plasma mass spectrometry (ICP-MS), which provides the required precision and sensitivity for accurate quantification of trace heavy metals in the leachate. This technique, as adopted by Aruta et al. (2022) and Datta et al. (2021) ensures high-accuracy analysis of trace elements in environmental samples.

To identify the CoC, leachable parameters were assessed against background levels and the prescribed permissible limits for disposal in engineered landfills and reuse in off-site applications. Parameters exceeding either the background levels or the permissible limits for disposal and reuse were classified as CoC. These contaminants require control to ensure the SLW is suitable for safe disposal or repurposing as a secondary resource, supporting sustainable techno-economic development.

To ensure reliable and reproducible results, all samples were analyzed in triplicates. Accurate calibration checks were performed at regular intervals, and concurrent blank specimens were used throughout the experimental process to validate the results. Additionally, analytical-grade chemicals and reagents were employed to minimize the risk of contamination, ensuring the chemical integrity of the test samples. A stringent cleaning protocol was followed for all glassware used in the experiments. Prior to analysis, glassware was soaked in diluted nitric acid solution, followed by thorough rinsing with DI water to decontaminate and remove any residual acids, thereby minimizing the risk of cross-contamination.

11.2.4 Field Study for Improved Quality of SLW

Field experiments aimed at improving the quality of the CF (30–6 mm) were conducted at the dumpsite. The necessary equipment and machinery were provided by the relevant authorities, while the field trommeling operation was facilitated by DCC Infra Pvt. Ltd., an agency involved in the bio-mining of legacy waste. The field experiment, termed Field Repeated Trommeling (FRT), was performed in duplicate (FRT-1 and FRT-2) on two separate dates to validate the consistency of the results. Each FRT cycle consisted of the initial trommeling (Tr-1), followed by three subsequent rounds of repeated trommeling (RT), labeled Tr-2, Tr-3, and Tr-4. These iterations were conducted to evaluate the effectiveness of FRT in improving the removal of FF from the CF. Figure 11.3 illustrates the sequence of FRT process adopted in the study. The current rapid field trommeling method, implemented on-site, was replicated to ensure the reproducibility of the results in future work. Legacy waste was excavated using a backhoe excavator, and the freshly excavated waste was fed into a hopper connected to a 30 mm trommel for segregation.

Large-sized materials, such as boulders and wooden pieces, were manually removed from the MLW before it was conveyed to the 30 mm trommel. The fractions finer than 30 mm were then transferred to a 6 mm trommel, which segregated the input material into two categories: the +6 mm fraction (30–6 mm, CF) and the −6 mm fraction (FF). To account for the initial volume of input material (−30 mm fractions) fed into the 6 mm trommel before the RT process, the two trommels (30 and 6 mm) were deliberately kept unconnected in sequence at the study site.

After the initial trommeling through the 6 mm trommel, the coarse fraction (30–6 mm) was quantified. The quantity of the input material (−30 mm fraction) and the coarse fraction was determined by manually counting the number of bucket loads using a backhoe loader. To assess the initial fines content and particle size distribution of the segregated fractions, laboratory wet sieve analysis was conducted. For enhanced removal of FF from the CF, the segregated coarse fraction was fed back into the 6 mm trommel, and this process was repeated three times. After

Fig. 11.3 FRT operation at one dumpsite: feeding of ETF to 6 mm trommel (**a**, **b**); CF segregated from ETF (**c**); FF segregated from ETF (**d**); Feeding of segregated CF to 6 mm trommel (**e**, **f**); CF after repeated trommeling (**g**)

each trommeling cycle, samples of both the FF and FF were collected for further laboratory sieve analysis. A laboratory-scale leaching study (SSBL) was carried out on the CF obtained from each trommeling cycle (Tr-1, Tr-2, Tr-3, and Tr-4) using DI water. The effects of repeated trommeling on the fines content in the coarse fraction and the leaching of contaminants were studied. Parameters such as pH, EC, release of colored leachate, and dissolved salts were measured during the batch leaching test. Moisture content and organic content of the segregated fractions from each trommeling repetition were also analyzed to observe variations. Overall, the quality of the coarse fraction was evaluated in terms of fines removal and the reduction of leachable contaminants, with an aim to determine its suitability as construction material.

11.3 Results and Discussions

11.3.1 Physical Characterization

Compositional analysis of MLW is crucial in understanding its heterogeneity, which can significantly impact its recycling and reuse potential. Identifying the predominant components within the waste stream allows for optimizing technological advancements to improve segregation, recycling, and recovery of valuable secondary resources. The present study, analyzing mixed MSW aged over 10–20 years, reveals that fractions finer than 20 mm—comprising primarily of FF and CF account for more than 60% of the total waste analyzed (see Fig. 11.4). This is largely due to the prolonged decomposition and degradation of waste over 10 to 20 years at the dumpsite. Additionally, the disposal of C&D debris has contributed approximately 20% to the total waste composition, which precedes the finer fractions in the MLW. The findings of this study align well with previous research on legacy MSW in India (Kurian et al. 2003; Somani et al. 2020), as well as studies conducted at Belgian dumpsites (Quaghebeur et al. 2013), Finnish dumpsites (Rong et al. 2017), and Chinese dumpsites (Sormunen et al. 2008). This comparison consistently reflects the global composition of mixed legacy MSW. However, combustible materials—such as plastics, corrugated paper, textiles, and wood—represent around 15% of the total waste, while non-combustible materials—like glass and metals—contribute less than 1% of the total MLW composition.

A comprehensive compositional analysis is essential for identifying the types and quantities of components that can be recovered from legacy waste. In this study, it was observed that the recovery and recycling of metals is not particularly promising due to their minimal presence in legacy waste. However, the recovery and

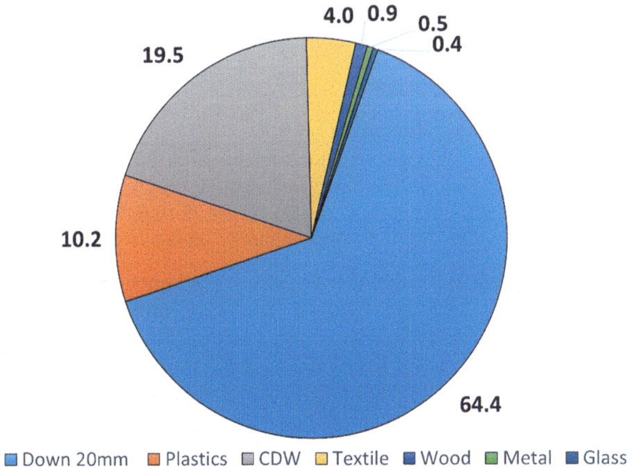

Fig. 11.4 Percentage content of components of mixed legacy waste

recycling of soil and gravel-like material as secondary resources for use in soil and construction materials show significant potential, given their substantial content in the legacy waste. Therefore, compositional analysis is a fundamental step in planning any resource recovery program from legacy waste. Additionally, site-specific studies are crucial for accurately estimating the recoverable and recyclable resources from the waste deposited at that particular site.

The grain size distribution of MLW, as presented in Table 11.1, reveals a significant presence of SLM finer than 4.75 mm, accounting for 60% by weight, followed by GLM in the size range of 31.5–4.75 mm, which constitutes 18.5%. Together, these two fractions make up nearly 80% of the total mixed legacy MSW. This highlights the substantial recovery potential of SLM and GLM from mixed legacy MSW.

The GSD analysis also highlighted the inefficiency of the ongoing on-site segregation by trommeling, which results in poor segregation of the CF. Specifically, a significant amount of FF is present in the CF, as shown in Fig. 11.5. Ideally, the CF (30–6 mm fraction) should not contain any fraction smaller than 6 mm (FF). However, the challenges associated with the rapid trommeling operation at the studied dumpsite were evident in the GSD analysis of the segregated fractions. These challenges, which lead to poor-quality segregated fractions, prompted the authors to investigate methods to improve the quality of segregated fractions and enhance their potential for recovery and reuse as construction materials in various off-site applications.

The moisture content of legacy waste being segregated by trommeling plays a crucial role in the quality of the segregated fractions. Table 11.2 presents the MC of the segregated legacy waste. The challenges in rapid trommeling are exacerbated by

Table 11.1 Grain size distribution of the legacy waste

Waste category		Particle size (mm)	Percent content (w/w)
Mixed legacy waste		+75	5.0
		75 to 31.5	16.5
		31.5 to 4.75	18.5
		4.75 to 0.075	38.3
		–0.075	21.7
Segregated legacy waste (SLW)	Entire trommeled fraction (−30 mm)	31.5 to 4.75	26.0
		4.75 to 0.075	46.8
		–0.075	29.4
	Coarse fraction (30–6 mm)	Low fines (−4.75 mm)	21.6
		High fines (−4.75 mm)	70.2
	Fine fraction (−6 mm)	4.75 to 0.075	74.3
		–0.075	25.7

Fig. 11.5 Appearance of CF before washing (**a**, **b**, **c**): (**a**) low content of FF, (**b**) and (**c**) high content of FF; after washing (**d**, **e**, **f**): (**d**) low FF content, (**e**) and (**f**) high FF

the significant moisture content in the legacy waste being processed at the dumpsite. With over 24% MC in the ETF, which is segregated into coarse and fine fractions, the trommeling efficiency is hindered. The moist waste tends to clog the trommel screen, closing off its perforations and reducing segregation effectiveness. Furthermore, the adherence of moist FF to the surface of the CF leads to a higher content of FF in the CF, as confirmed by the GSD analysis

However, due to the prevailing field conditions, operational challenges, and the urgent need to remove large quantities of legacy waste, it becomes difficult for the operating contractors to implement the time-consuming processes of pre-drying and other measures as outlined in the legacy waste management guidelines (CPCB 2019). These challenges hinder the adoption of the ideal approach for LFM. The objectives of the present study are aligned with addressing these concerns related to the rapid segregation of legacy waste.

Table 11.2 Physicochemical constituents of segregated legacy waste

Parameters	Unit	30 to 6 mm (CF)	– 30 mm (ETF)	– 6 mm (FF)	Background soil (Delhi Silt)
M.C	%	11.8 ± 2.1	24.3 ± 2.8	27.7 ± 0.6	1.8–2.5
O.C	%	11.7 ± 1.4	12.0 ± 1.2	12.3 ± 1.3	0.9–1.2
pH	–	7.8 ± 0.339	7.9 ± 0.168	7.7 ± 0.076	7.1–7.2
E.C	mS/cm	1.148 ± 0.221	1.708 ± 0.202	2.482 ± 0.237	0.20–0.32
Color	PCU	156.7 ± 24.8	251.7 ± 72.6	238.3 ± 21.1	25–30
TDS	mg/kg	10,420 ± 3285	16,083 ± 2771	22,213 ± 2782	500–700
Sulphates		2421 ± 478	4467 ± 1104	7239 ± 1016	300–350
Chlorides		2119 ± 530	2531 ± 216	3529 ± 305	200–250
Sodium		632.2 ± 90.5	1748 ± 794	2536 ± 1160	80–100
Magnesium		171 ± 48	359 ± 73	457 ± 69	40–60
Potassium		414 ± 65	965 ± 192	1234 ± 256	30–70
Calcium		151 ± 43	259 ± 56	299 ± 55	120–180
Arsenic	mg/kg	0.12 ± 0.02	0.14 ± 0.06	0.17 ± 0.06	0.04–0.06
Chromium		0.04 ± 0.01	0.04 ± 0.01	0.04 ± 0.01	0.02
Copper		0.56 ± 0.27	0.52 ± 0.15	0.44 ± 0.04	0.25–0.52
Nickle		0.16 ± 0.04	0.22 ± 0.12	0.22 ± 0.07	0.13–0.21
Lead		0.02 ± 0.02	n.d	n.d	0.03–0.06
Zinc		0.57 ± 0.28	0.38 ± 0.14	0.51 ± 0.14	0.22–0.67

11.3.2 Chemical Characterization and Contaminants of Concern (CoC)

Although the present study primarily focuses on the CF of the SLW, it is important to understand the broader contamination potential of the SLW. For this purpose, the chemical parameters of the ETF (−30 mm), CF (30–6 mm), and FF (−6 mm) were analyzed. A detailed discussion on the chemical constituents and their interactions across different size fractions (ETF, CF, FF), along with various leaching procedures including SSBL and Up-Flow Column Leaching (UCL), is provided in the earlier published research by the authors, Parida et al. (2024), which precedes the current study. A summary of the chemical constituents obtained through SSBL of the segregated fractions (ETF, CF, FF) is presented in Table 11.2.

It can be observed that FF has a higher affinity for contaminants compared to CF when analyzing the concentration of contaminants across different size fractions of SLW. Furthermore, particle size analysis reveals that the poorly segregated CF contains a significant amount of FF. This indicates that the high content of FF is the primary factor contributing to the elevated contamination levels in the CF. As given in Table 11.2, the leachable heavy metals in the CF are below background levels, suggesting a low contamination potential from heavy metals leaching. However, the leaching of dissolved salts is considerably higher than the background levels, which limits the potential for CF reuse as an alternative material in off-site applications. Due to the elevated salt content, electrical conductivity—an indicator of dissolved ion concentrations—is notably high in the SLW fractions. Consequently, it can be concluded that dissolved salts are the major CoC in the CF.

This study examines the impact of the proposed field remedial measures on the CoC and assesses the suitability of the improved segregated waste (CF) for reuse in various off-site applications.

11.3.3 Evaluation of FRT for Improved Quality of SLW

This section addresses the impact of FRT on enhancing FF removal from CF, subsequently reducing the levels of contaminants of concern within the segregated CF, in contrast to the conventional trommeling method commonly practiced in India. This section delves into the parametric variations observed in the CF due to RT conducted at the dumpsite.

11.3.3.1 Effect on M.C and O.C

The initial M.C and O.C of the ETF, CF, and FF were determined in the laboratory. Subsequently, the M.C and O.C of the CF were assessed for each repetition of trommeling during the FRT process, as presented in Table 11.3. Initially, the M.C of FF and ETF were notably higher than that of the CF, which can be attributed to the immediate trommeling of legacy waste post-excavation, without prior drying and stabilization. The lower M.C observed in CF is likely due to the presence of gravel-like materials. Throughout the FRT process, the M.C of CF consistently decreased with each repetition of trommeling. For instance, the initial M.C of 11.8 ± 2.1% reduced to 5.9 ± 1.2% after three repetitions of trommeling, resulting a 50% reduction in M.C of CF during the FRT process. This reduction can be attributed to continuous handling of CF and exposure to air, promoting aeration during the FRT process.

The O.C of FF and ETF exceeded that of CF. Notably, a significant reduction in O.C of CF was observed during the FRT process, with O.C decreasing from 11.7 ± 1.4% to 5.3 ± 0.8% after multiple repetitions of trommeling, representing a substantial reduction of 55%. This reduction in O.C can be linked to the diminished presence of FF within CF during the FRT process, as evident by the initial O.C of FF (12.3 ± 0.3%) observed during the first trommeling.

Table 11.3 Effect of FRT process on M.C and O.C of CF

RT Phases	M.C (%)	O.C (%)
Tr-1	11.8 ± 2.1	11.7 ± 1.4
Tr-2	8.4 ± 0.9	9.6 ± 1.8
Tr-3	7.0 ± 0.8	7.5 ± 1.8
Tr-4	5.9 ± 1.2	5.3 ± 0.8
% change	50.00	54.70

Table 11.4 Effect of FRT on removal of FF from CF

Trommeling phase	M.C (%)	FRT-1 Removal of FF from CF by field trommeling (%)	FRT-1 FF content in CF by laboratory sieving (%)	FRT-2 Removal of FF from CF by Field trommeling (%)	FRT-2 FF content in CF by laboratory sieving (%)
Tr-1	11.8 ± 2.1	–	24.5	–	21.6
Tr-2	8.4 ± 0.9	16.7	12.8	15	8.9
Tr-3	7.0 ± 0.8	6.7	11.6	7.5	7.0
Tr-4	5.9 ± 1.2	3.3	9.8	3.8	6.7

11.3.3.2 Effect on Fine Fraction

The investigation focused on elucidating the impact of FRT on the efficacy of fine fraction (−6 mm) removal from the CF. During the meticulous FRT process, precise quantification of input bucket loads of CF was undertaken, facilitating the calculation of FF removal by delineating the difference between the consecutive input bucket loads. This rigorous methodology ensured accuracy and consistency throughout the FRT processes. CF samples obtained at each RT stage underwent rigorous laboratory GSD analysis, supplemented by meticulous washed sieving to ensure precise determination of FF presence in CF.

A progressive decline in overall FF removal was observed with the advancement of RT phases (see Table 11.4), indicating a nuanced interplay between trommeling dynamics and moisture content. During the initial RT stage of CF, FF percentages were notably lower at 16.7% and 15.0% in FRT-1 and FRT-2, respectively, compared to corresponding laboratory washed sieve analyses, which yielded 24.5% and 21.6% for FRT-1 and FRT-2, respectively. This discrepancy underscores a potential reduction in the efficacy of trommeling operations, likely stemming from factors such as elevated moisture content, trommel screen clogging, and the dampened surface characteristics of CF during onsite trommeling endeavors.

This meticulous analysis unveils critical insights into the intricate dynamics of trommeling operations, offering valuable implications for optimizing FF removal efficiency in dumpsite settings.

11.3.3.3 Effect on Leaching Parameters

To comprehensively assess the impact of RT on the leaching characteristics of the CF, laboratory single-stage batch leaching procedure was conducted. The SSBL test was executed at a L/S ratio of 10 L/kg. The leaching parameters analyzed for CF derived from the FRT process encompassed pH, EC, color intensity, cations (Na, Mg, K, Ca), anions (sulphates, chlorides), total dissolved solids (TDS) and trace metals (As, Cr, Cu, Ni, Pb, Zn). Herein, the influence of FRT on the leaching parameters of CF is discussed.

Table 11.5 represents the SSBL test results corresponding to each phase of trommeling within FRT-1 and FRT-2. From these results, several salient inferences can be gleaned, shedding light on the efficacy of the trommeling process and its ramifications on leachability characteristics.

Table 11.5 Effect of FRT process on leachable parameters from SSBL test (at L/S = 10 L/kg)

S. No.	Parameter	Unit	Tr-1	Tr-2	Tr-3	Tr-4	% change in concentration
1	pH	–	7.8 ± 0.3	7.9 ± 0.2	8.0 ± 0.0	8.0 ± 0.1	–
2	EC	mS/cm	1.307 ± 0.196	1.063 ± 0.297	0.872 ± 0.098	0.707 ± 0.142	45.88
3	Color	PCU	156.6 ± 24.8	120.5 ± 22.4	85.3 ± 19.8	70.5 ± 11.1	55.00
	Dissolved solids						
4	TDS	mg/kg	10,420 ± 3285.4	7526.7 ± 2911.2	6840.0 ± 802.3	5163.3 ± 984.9	50.45
5	Sulphates		2421.3 ± 313.2	1760.2 ± 134.9	1602.1 ± 246.1	1526.5 ± 268.8	36.95
6	Chlorides		2119.3 ± 530.7	1813.3 ± 169.1	1555.1 ± 130.6	1509.5 ± 337.1	28.77
7	Sodium		632.2 ± 90.5	499.2 ± 73.6	440.6 ± 86.6	402.0 ± 102.4	36.41
8	Magnesium		171.7 ± 48.9	108.2 ± 9.3	94.0 ± 15.9	78.4 ± 11.6	54.30
9	Potassium		414.4 ± 65.8	371.2 ± 42.3	353.0 ± 71.5	319.8 ± 83.3	22.82
10	Calcium		151.9 ± 43.2	104.3 ± 15.4	89.8 ± 19.0	72.2 ± 18.8	52.47
	Trace metals						
11	Arsenic (as)	mg/kg	0.126 ± 0.022	0.112 ± 0.010	0.080 ± 0.015	0.078 ± 0.032	38.03
12	Chromium (Cr)		0.056 ± 0.013	0.046 ± 0.009	0.044 ± 0.004	0.028 ± 0.014	50.42
13	Copper (cu)		0.624 ± 0.555	0.592 ± 0.346	0.408 ± 077	0.230 ± 0.088	63.09
14	Nickle (Ni)		0.144 ± 0.028	0.125 ± 0.047	0.122 ± 0.031	0.122 ± 0.060	15.52
15	Lead (Pb)		0.159 ± 0.011	0.035 ± 0.008	0.025 ± 0.004	0.011 ± 0.000	92.97
16	Zinc (Zn)		0.636 ± 0.082	0.550 ± 0.152	0.300 ± 0.028	0.252 ± 0.048	60.38

pH, E.C, Color Intensity

The pH of the segregated legacy waste demonstrates a narrow range (7.6–7.9), indicative of a stable and significantly degraded waste profile. Specifically, the mean pH value of the CF was determined to be 7.8. Notably, this trend persisted during the RT of CF, with the mean pH of CF maintaining a range-bound consistency (7.8–8.0) throughout the FRT process. From these observations, it can be inferred that the pH remains largely unaffected by the successive RT iterations applied to CF during the two FRT processes.

The E.C of the CF experienced substantial alterations due to the FRT process. Initially, the mean E.C of 1.307 mS/cm experienced a notable reduction to 0.707 mS/cm, marking an overall cumulative decrease of 45% following the removal of FF through repeated trommeling. At each stage of the RT, an average reduction of 18% was observed. The presence of elevated E.C levels signifies a notable concentration of dissolved ions in the soil and water matrix. Consequently, the reduction in E.C holds the potential to mitigate surface and groundwater contamination attributed to the leaching of dissolved ions.

The color intensity of the water extract, quantified in PCU, exhibited a noteworthy reduction, amounting to less than half of the initial intensity. Preceding RT, the color intensity stood at 156.7 PCU, witnessing a reduction of 55% to 70.5 PCU after three RT cycles of the CF. As delineated in Chap. 5, heightened color intensity (ranging from 238 to 285 PCU) is indicative of elevated FF content, attributable to the presence of organic matter. The substantial decline in color intensity observed in CF can be attributed to FF reduction facilitated by the FRT process.

The reduction rate in the initial two RT cycles was recorded at 23.1% and 29.2%, respectively, subsequently diminishing to 17.4% in the third RT cycle, underscoring the decreased FF presence in CF during the later stages of the FRT process.

Dissolved Solids

The TDS content, reflective of dissolved ions, initially registered at 10420 mg/kg before RT. Following three RT cycles of segregated CF, the TDS content exhibited a reduction exceeding 50%, plummeting to 5163 mg/kg. This substantial decrease in TDS content, facilitated by the FRT process, suggests a potential reduction in individual dissolved solids within the segregated CF.

Among the analyzed dissolved solids, water-soluble sulphates and chlorides emerged as predominant parameters, with concentrations ranging from 1960 to 2792 mg/kg and 1582 to 2715 mg/kg, respectively, in the segregated CF before the FRT process. The reduction rates of these constituents were particularly notable during the initial RT cycle, reaching 27.3% for sulphates and 14.4% for chlorides. However, by the third RT cycle, these reduction rates dwindled to less than 5%. The cumulative removal percentages for sulphates and chlorides stood at 37% and 30%, respectively. Additionally, minor dissolved solids like Mg and Ca experienced significant reductions of 54.3% and 52.5%, respectively, while Na and K were reduced by 36.4% and 22.8%, respectively, after three RT cycles of CF. These observations underscore the comprehensive impact of the FRT process on dissolved solids, primarily attributed to the reduction of fine fraction content in CF.

Trace Metals

The impact of FRT on the trace metal composition in the segregated CF exhibited notable variations. Reduction levels ranged significantly, with Ni experiencing a modest 15.5% decrease, while Pb demonstrated the highest sensitivity with an impressive 93% reduction. Moreover, notable reductions exceeding 50% were observed for Cr, Cu, and Zn, while As exhibited a moderate 38% decline following FRT. The substantial decrease in FF content directly corresponded to a marked reduction in trace metal concentration, yielding a cleaner CF. However, the implications of this cleaner CF are contingent upon adherence to regulatory thresholds for its utilization in construction materials, an aspect thoroughly investigated in subsequent section.

11.3.4 Assessment of CF for Bulk Reuse Options

The leachable parameters were observed to be significantly influenced by the FRT process where the dissolved solids including sulphates and chlorides were reduced by 22–54% and heavy metals by 15% to as high as 92% indicating the effectiveness of the FRT process in improving the quality of CF. however, it becomes necessary to assess the improved quality of CF against the prescribed regulatory limits for off-site reuse applications. The reuse criteria encompassing open reuse, reuse with surface sealing and cohesive soil layers and isolated reuse options prescribed in the international regulations were assessed and are summarized in Table 11.6.

Despite of four stages of RT of CF, the contents of dissolved solids are significantly high, in the range of 5 to 10 times compared to that of the background level. This indicates any unrestricted use of such material can elevate the contaminant concentration in the existing soils and leach to the surrounding environment contaminating the surface and subsurface water sources. The content of heavy metals are comparable and marginally higher than the background levels that indicate that the contamination by the leaching of heavy metals by the unrestricted use of CF obtained after the FRT process is unlikely. This demonstrates that the FRT process is effective in reducing the content of leachable heavy metals to the background level.

Comparing the regulatory limits for reuse, it is evident that the significant content of dissolved solids-sulphates and chlorides limits the reuse of improved CF for unrestricted and open reuse in off-site applications prescribed in the German and the Dutch standards, respectively. However, as per the regulatory limits, the improved CF can be used in isolated environment or with the provision of surface sealings to restrict the leaching of contaminants to safe keep the immediate environment in the

Table 11.6 Comparison of leachable parameters of improved CF with the international regulatory limits

Description		Dissolved Solids (mg/kg)			Leachable trace metals (mg/kg)					
		TDS	Sulphates	Chlorides	As	Cr	Cu	Ni	Pb	Zn
FRT-1&2	Improved CF from Tr-4	5163.3 ± 984.9	1526.5 ± 268.8	1509.5 ± 337.1	0.078 ± 0.032	0.028 ± 0.014	0.230 ± 0.088	0.122 ± 0.060	0.011 ± 0.0	0.252 ± 0.048
Background level	Delhi silt	500–700	300–350	200–250	0.04–0.06	0.02	0.25–0.52	0.13–0.21	0.03–0.06	0.22–0.67
International regulations	Applications									
German standard (LAGA 2012)	No restrictions at all	n.a	500	100	0.1	0.15	0.5	0.4	0.2	1
	No sealing		1500	200	0.1	0.3	0.5	0.5	0.4	1
	Cohesive soil layer between material and GW		3000	400	0.4	0.75	1.5	1	1	3
	Sealing of surface		6000	1500	0.5	1	2	1	1	4
Dutch standard (VROM 2007)[a]	Open reuse		1730	616	0.9	0.63	0.9	0.44	2.3	4.5
	Isolated reuse		20,000	8800	2	7	1	2.1	8.3	14

Note: [a]adopted from Soil Quality Decree (2007); n.a- not available; n.d- not detected; FRT-Field Repeated Trommeling; CF-Coarse Fraction; Tr-4-Trommeling stage 4; GW-Groundwater

long run. Since the leachable heavy metals are significantly reduced by FRT technique, they do not limit the unrestricted or open reuse of improved CF.

It is conclusive that the FRT technique has significantly influenced in improving the quality of CF, however this could not yield results that can allow the unrestricted or open reuse of CF in off-site reuse applications replacing the natural resources.

11.4 Summary and Conclusions

Developing an onsite full-scale methodology that is both simple and unique presents an opportunity to gain insights into real-world scenarios concerning the management of contaminants in segregated legacy MSW. By assessing the effectiveness of field-scale measures in reducing contaminant concentrations in SLW, municipal authorities and policymakers can consider adopting similar methodologies for legacy MSW management. The methodology employed in this study holds promise for alleviating the burden of contaminant concentration in segregated CF prior to its use in offsite applications and as construction material.

The key findings of the study are summarized below.

- A proportional reduction in both the M.C and O.C was observed following the repeated trommeling of CF. This reduction can be attributed to the continuous handling of CF, which promotes aeration and drying, thereby leading to a decrease in the fine fraction during the FRT process.
- The considerable presence of fine fraction within the segregated CF can be attributed to the factors such as elevated moisture content of the excavated legacy waste, clogging of trommel screen, and the adhesive nature of CF surfaces, facilitating fine fraction adhesion. Subsequently, the implementation of the FRT process led to a noteworthy reduction in the fine fraction.
- The pH levels of water extracts from CF samples collected at both Tr-1 and Tr-4 remained unaffected by the leaching methods, L/S ratios and the FRT process.
- The leaching parameters, encompassing E.C, color intensity, dissolved solids, and trace metals, were notably influenced by both the L/S ratio and the FRT process.

The ultimate fate of the CF obtained at the end of the FRT process hinges upon adherence to regulatory threshold limits governing contaminant concentration for its viable reuse as soil and construction materials in off-site applications.

References

Abdel-Shafy HI, Mansour MSM (2018) Solid waste issue: sources, composition, disposal, recycling, and valorization. Egypt J Pet 27(4):1275–1290. https://doi.org/10.1016/j.ejpe.2018.07.003

APHA/AWWA/WEF (2012) Standard methods for the examination of water and wastewater, 23rd edn. American Public Health Association, American Water Works Association, Water Environment Federation, Denver

Aruta A, Albanese S, Daniele L, Cannatelli C, Buscher JT, De Vivo B et al (2022) A new approach to assess the degree of contamination and determine sources and risks related to PTEs in an urban environment: the case study of Santiago (Chile). Environ Geochem Health 6. https://doi.org/10.1007/s10653-021-01185-6

ASTM D2974 (2014) Standard test methods for moisture, ash, and organic matter of peat and other. ASTM Int 1–4. https://doi.org/10.1520/D2974-14.obtaining

ASTM D5231 (2016) Standard test method for determination of the composition of unprocessed municipal solid waste https://doi.org/10.1520/D5231-92R16.Copyright

ASTM D6913 (2017) Standard test methods for particle-size distribution (gradation) of soils using sieve. ASTM International 04. https://doi.org/10.1520/D6913

Burlakovs J, Kaczala F, Vincevica-Gaile Z, Rudovica V, Orupõld K, Stapkevica M et al (2016) Mobility of metals and valorization of sorted fine fraction of waste after landfill excavation. Waste Biomass Valor 7(3):593–602. https://doi.org/10.1007/s12649-016-9478-4

CPCB (2019) Guidelines for disposal of legacy waste. Central Pollution Control Board (CPCB), India

Danthurebandara M (2016) Environmental and economic performance of enhanced landfill mining., (January 2015)

Datta M, Somani M, Ramana GV, Sreekrishnan TR (2021) Feasibility of re-using soil-like material obtained from mining of old MSW dumps as an earth-fill and as compost. Process Saf Environ Prot 147:477–487. https://doi.org/10.1016/j.psep.2020.09.051

De Vocht AJPJP, Descamps S (2014) Biodiversity and Enhanced Landfill Mining : Weighting local and global impacts. In: International Academic Symposium on Enhanced Landfill Mining (November), pp 275–290. https://doi.org/10.13140/2.1.4969.8245

Esguerra J, Sauve G, Einhäupl P, Krook J (2020) European training network for resource recovery through enhanced landfill mining (NEW-MINE) D4.1 final report on WP multi-criteria assessment, (721185), 1–8. Retrieved from http://new-mine.eu/

Frändegård P, Krook J, Svensson N (2015) Integrating remediation and resource recovery: on the economic conditions of landfill mining. Waste Manag 42:137–147. https://doi.org/10.1016/j.wasman.2015.04.008

Hogland W, Marques M (2010) Enhanced landfill mining: material recovery, energy utilization and economics in the EU (Directive) perspective. In: Enhanced landfill mining and the transition of sustainable materials management: proceedings of the international academic symposium on enhanced landfill mining, Houthalen-Helchteren, Belgium (January), pp 209–222. Retrieved from http://lnu.diva-portal.org/smash/record.jsf?pid=diva2:426824

Hölzle I (2018) Dry screening—assessing the effectiveness of contaminant reduction in recovered landfill soils. J Clean Prod 172:1998–2008. https://doi.org/10.1016/j.jclepro.2017.11.225

Hull RM, Krogmann U, Strom PF (2005) Composition and characteristics of excavated materials from a New Jersey landfill. J Environ Eng 131(3):478–490. https://doi.org/10.1061/(asce)0733-9372(2005)131:3(478)

Jani Y, Kaczala F, Marchand C, Hogland M, Kriipsalu M, Hogland W, Kihl A (2016) Characterisation of excavated fine fraction and waste composition from a Swedish landfill. Waste Manag Res 34(12):1292–1299. https://doi.org/10.1177/0734242X16670000

Jones PT, Geysen D, Tielemans Y, Van Passel S, Pontikes Y, Blanpain B et al (2013) Enhanced landfill mining in view of multiple resource recovery: a critical review. J Clean Prod 55:45–55. https://doi.org/10.1016/j.jclepro.2012.05.021

Krook J (2021, 23) Enhanced landfill mining, concept and challenges. Field Actions Sci Rep:20–21. Retrieved from https://journals.openedition.org/factsreports/6549

Krook J, Baas L (2013) Getting serious about mining the technosphere: a review of recent landfill mining and urban mining research. J Clean Prod 55:1–9. https://doi.org/10.1016/j.jclepro.2013.04.043

Kurian J, Esakku S, Palanivelu K, Selvam A (2003) Studies on landfill mining at solid waste dumpsites in India. In: Ninth international waste management and landfill symposium, vol 1994, pp 6–10

LAGA (2012) Requirements for the material recycling of mineral residues/waste (Anforderung an Die Stoffliche Ver-wertung Von Mineralischen Reststoffen/Abfällen). Stand. Erich Schmidt Verlag, Berlin

Lama N (2024) The World has a Waste Problem. Retrieved from https://www.ifc.org/en/blogs/2024/the-world-has-a-waste-problem#

Ministry of Housing and Urban Affairs (2025) Status of legacy waste management. Retrieved January 28, 2025, from https://sbmurban.org/swachh-bharat-mission-progess

Municipal Corporation of Delhi (2024) Solid waste management in MCD. Retrieved from http://mcdonline.gov.in/tri/ndmc_mcdportal/dems/

Parida D, Ramana GV, Datta M (2024) Investigation on trommeled legacy waste from full-scale mining of old dumpsites: suitable for valorization or scientific disposal? J Environ Manag 356(October 2023):120580. https://doi.org/10.1016/j.jenvman.2024.120580

Parrodi JCH, López CG, Küppers B, Raulf K, Vollprecht D, Pretz T, Pomberger R (2019) Case study on enhanced landfill mining at mont-saintguibert landfill in Belgium: characterization and potential of fine fractions. Detritus 8(December):47–61. https://doi.org/10.31025/2611-4135/2019.13877

Quaghebeur M, Laenen B, Geysen D, Nielsen P, Pontikes Y, Van Gerven T, Spooren J (2013) Characterization of landfilled materials: screening of the enhanced landfill mining potential. J Clean Prod 55:72–83. https://doi.org/10.1016/j.jclepro.2012.06.012

Rong L, Zhang C, Jin D, Dai Z (2017) Assessment of the potential utilization of municipal solid waste from a closed irregular landfill. J Clean Prod 142:413–419. https://doi.org/10.1016/j.jclepro.2015.10.050

Siddiqua A, Hahladakis JN, Al-Attiya WAKA (2022) An overview of the environmental pollution and health effects associated with waste landfilling and open dumping. Environ Sci Pollut Res 29(39):58514–58536. https://doi.org/10.1007/s11356-022-21578-z

Somani M, Datta M, Ramana GV, Sreekrishnan TR (2020) Contaminants in soil-like material recovered by landfill mining from five old dumps in India. Process Saf Environ Prot 137:82–92. https://doi.org/10.1016/j.psep.2020.02.010

Sormunen K, Ettala M, Rintala J (2008) Detailed internal characterisation of two Finnish landfills by waste sampling. Waste Manag 28(1):151–163. https://doi.org/10.1016/j.wasman.2007.01.003

SS-EN 12457-2 (2002) Characterization of waste-leaching-compliance test for leaching of granular waste materials and sludges. Part 2: one stage batch test at a liquid to solid ratio of 10 L/kg for materials with particle size below 4mm (without or with size reduction). Swedish Standard Institute, Stockholm

SS-EN 12457-4 (2002) Characterisation of waste leaching compliance test for leaching of granular waste materials and sludges Part 4: one stage batch test at a liquid to solid ratio of 10 l/kg for materials with particle size below 10 mm (without or with size reduction). Swedish Standard Institute, Stockholm

United Nations Environment Programme (2024) Beyond an age of waste—turning rubbish into a resource. Global Waste Management Outlook 2024. Retrieved from https://wedocs.unep.org/20.500.11822/44939

Vinti G, Bauza V, Clasen T, Medlicott K, Tudor T, Zurbrügg C, Vaccari M (2021) Municipal solid waste management and adverse health outcomes: a systematic review. Int J Environ Res Public Health 18(8):1–26. https://doi.org/10.3390/ijerph18084331

VROM (2007) Soil quality decree. Spatial Planning and the Environment, Ministry of Housing, Spatial Planning and the Environment, The Hague (Regeling Bodemkwaliteit., Ruimte en Milieu, Ministerie van Volkshuisvesting, Ruimtelijke Ordeling en Milieubeheer, Den Haag)

Waste Atlas (2014) The world's 50 biggest dumpsites. J Chem Inf Model 53

Wilson D, Rodic L, Modak P, Soss R, Rogero A, Velis C, ... Simonett O (2022) Global waste management outlook

Zhang L, Zhong Y, Geng Y (2019) A bibliometric and visual study on urban mining. J Clean Prod 239:118067. https://doi.org/10.1016/j.jclepro.2019.118067

Part III
Combustible Fraction

Chapter 12
Valorisation Potential of Combustible Fraction of Legacy Waste Obtained from Landfill Mining

Roshan Vilasrao Mankhair and Munish Kumar Chandel

12.1 Introduction

The waste management sector is currently facing significant challenges largely due to rapid industrialization, urbanization, and increasing consumption patterns coupled with past waste management practices. Historically, landfilling (often involving open dumping and unscientific landfilling) was the most prominent practice of waste management. This uncontrolled disposal of waste now poses significant risks to the environment as well as to the human health. In the past decade, landfill mining has gained traction globally which is driven by increasing land scarcity in urban areas, stringent environmental regulations, and transitioning toward a more circular economy (Singh et al. 2022). Landfill mining has emerged as a sustainable waste management strategy, offering the potential for resource and land recovery along with environmental remediation (Quaghebeur et al. 2013).

Combustible waste is one of the major components of legacy waste recovered during the process of landfill mining, and its effective valorization is crucial for the success of a landfill mining project (Mankhair and Chandel 2024). This fraction typically consists of materials like plastic, textile, wood, and paper, with plastic being the predominant contributor (Jain et al. 2023). However, unlike fresh waste, the combustible fraction of legacy waste is often heavily contaminated. This is due to the prolonged exposure to landfill conditions with high moisture which causes impurities such as dirt (soil-like material) and organic residues to adhere to it, making its separation and cleaning difficult. As a result, recycling and waste-to-material (WtM) are typically not a recommended valorization options for this fraction. The

R. V. Mankhair · M. K. Chandel (✉)
Environmental Science and Engineering Department, Indian Institute of Technology Bombay, Powai, Mumbai, Maharashtra, India
e-mail: munish.chandel@iitb.ac.in

M. Somani et al. (eds.), *Landfill Mining of Legacy MSW: Global Insights and Indian Experience*, Springer Transactions in Civil and Environmental Engineering, https://doi.org/10.1007/978-981-95-3959-8_12

extensive cleaning required to make it suitable for recycling is an uneconomical option, while the resulting product still might be of inadequate quality to be used for recycling (Singh and Chandel 2020; Zhou et al. 2014).

Despite the challenges associated with the recycling of combustible fraction of legacy waste, it offers the potential to recover energy due to high calorific values. Therefore, it can be a suitable feedstock for processing into refuse-derived fuel (RDF) and waste-to-energy (WtE) applications (Mankhair and Chandel 2024). However, to determine its suitability for such applications, a thorough physico-chemical characterization is essential. This analysis provides the details of different parameters including moisture content, volatile matter, ash content, fixed carbon, and calorific value, which directly influence the combustion efficiency and in turn the energy recovery potential. For instance, high moisture and ash content reduce the calorific value of the waste, affecting the feasibility of its use as a feedstock for WtE treatment technologies. Knowledge of the volatile matter and fixed carbon content are important factors in optimizing the combustion processes for enhancing overall energy recovery. Based on the physicochemical characteristics of the combustible fraction of legacy waste, several studies have suggested WtE as the most viable valorization route for energy recovery (Bhatnagar et al. 2017; Bosmans et al. 2013; Cheela et al. 2021; Mankhair and Chandel 2024).

The current chapter investigates the physicochemical characteristics of the combustible fraction of legacy waste based on the findings from the available literature. By synthesizing research findings from both global and Indian studies, the chapter aims to provide a comprehensive understanding of the valorization potential of the combustible fraction of legacy waste. The focus is on its potential use as a feedstock for WtE applications, considering its high calorific value and suitability for energy recovery. Additionally, a SWOT (strengths, weaknesses, opportunities, and threats) analysis is performed to examine the key factors influencing the utilization of this fraction as fuel in WtE applications.

12.2 Composition of the Combustible Fraction of Legacy Waste

A study by Jain et al. (2023) reviewed 27 global studies covering 34 landfills and found that the combustible fraction comprises an average of 24.5% of the total legacy waste. At several landfill sites, the combustible fraction exceeds over 50% of the total legacy waste, making its effective valorization essential for the success of landfill mining projects. Among the combustible fraction, plastic contributes the largest share, followed by textile, wood, and paper (Fig. 12.1). Plastic waste accounts for approximately 12–28% of the total legacy waste, highlighting the importance of its efficient management and exploring potential resource recovery pathways (Cheela et al. 2021; Singh and Chandel 2020; Mankhair and Chandel 2024).

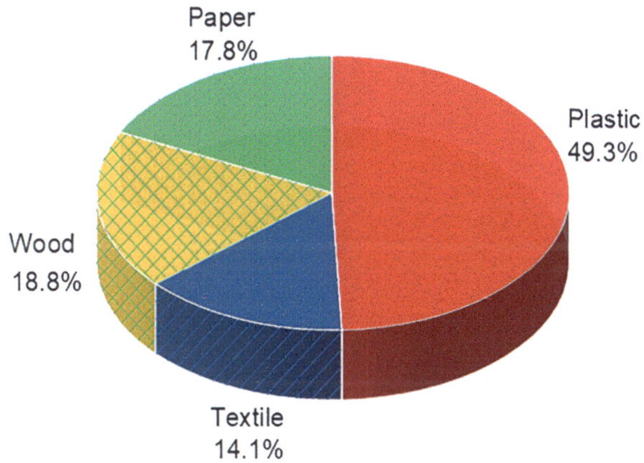

Fig. 12.1 Composition of the combustible fraction of legacy waste based on 27 global studies (Jain et al. 2023)

The percentage of combustible waste significantly varies across different countries, and it ranged from 4.3% to 54.9% of the total legacy waste in the referred studies (Tables 12.1 and 12.2). A study by Singh and Chandel (2020) observed that the overall percentage of combustible fraction followed a decreasing trend with increasing landfill age. This significant variation in the composition of legacy waste can be attributed to several factors, including consumption patterns, waste segregation practices, the type of waste being disposed of, and the age of the landfill (Singh et al. 2022). The composition of legacy waste varies drastically within and between landfills, and hence conducting a thorough compositional analysis is crucial for designing effective site-specific resource recovery strategies. The variations in the waste composition also alter the physical and chemical characteristics of the waste, further emphasizing the need for site-specific investigations.

The available literature studies have reported a very low amount of paper in the Indian legacy waste, ranging from 0.4% to 6.1%. This is primarily attributed to the lower amount of paper waste generated and disposed of in landfills, coupled with its decomposition over time. Several researchers have investigated the changes in the composition of waste fractions relative to landfill age (Kaartinen et al. 2013; Singh and Chandel 2020). They observed that paper was present in smaller quantities in older waste, indicating the change in consumption patterns over the years and the effects of prolonged decomposition. The studies outside India reported significantly higher proportions of paper in the combustible fraction of legacy waste. For instance, in Belgium, paper content reached up to 30% (Table 12.2). These differences can be attributed to varying waste composition, waste disposal practices, and the extent of decomposition in various regions. These findings emphasize the influence of socio-economic and environmental factors on the composition of legacy and underscore the need to consider regional differences in waste composition and characteristics when planning for landfill mining strategies.

Table 12.1 Composition of the combustible fraction of legacy waste in India

Component	Composition (%)						
Plastic	54.2	50.7	75.8	56.9	76.7	48.2	61.7
Textile	18.8	26.7	5.7	35.3	18.6	25.0	16.8
Wood	20.9	17.3	18.1	–	4.7	26.8	21.7
Paper	6.1	5.3	0.4	7.8	–	–	–
Overall combustible fraction (%)	52.1	22.9	16.8	29.5	4.3	5.6	6.0
Location	Pune	Deonar, Mumbai	Mulund, Mumbai	Visakhapatnam	Okhla, Delhi	Jawaharnagar, Hyderabad	Ukkayyapalli, Kadapa
Reference	Mankhair and Chandel (2024)	Mankhair et al. (2024)	Singh and Chandel (2020)	Cheela et al. (2021)	Somani et al. (2018)		

Table 12.2 Composition of the combustible fraction of legacy waste in other countries

Component	Composition (%)					
Plastic	78.8	54.1	69.6	51.2	77.0	63.0
Textile	5.9	6.7	–	16.3	15.3	31.2
Wood	–	8.9	14.6	15.8	1.3	5.2
Paper	15.3	30.3	15.8	16.7	6.4	0.6
Overall combustible fraction (%)	11.8	46.2	32.2	44.9	54.9	51.6
Location	Nanqijia, China	Remo, Belgium	Kudjape, Estonia	Kupio, Finland	Mont-saint-Guibert, Belgium	Tianjin, China
Reference	Prechthai et al. (2008)	Rong et al. (2017)	Quaghebeur et al. (2013)	Bhatnagar et al. (2017)	García López et al. (2019)	Tao et al. (2024)

12.3 Characteristics of the Combustible Fraction of Legacy Waste

The physicochemical characteristics of legacy waste play a critical role in the selection of appropriate valorization pathways. Key parameters to be considered for evaluating the suitability of combustible fraction for various applications include proximate analysis (moisture content, volatile solids, fixed carbon, and ash content), ultimate analysis [carbon (C), hydrogen (H) nitrogen (N), sulfur (S), and oxygen (O)], total organic carbon (TOC), calorific value, heavy metals, and chlorine content. For WtE applications, parameters such as calorific value, TOC, proximate analysis parameters, total C, H, and N are essential for evaluating the energy recovery potential. Additionally, the levels of sulfur and chlorine content are important parameters to estimate the levels of emissions during the thermal treatment processes. For valorization as WtM, the inorganic composition, particularly the heavy metal content is a critical factor to ensure the feasibility of material recovery (Quaghebeur et al. 2013).

Limited studies focussing on the thorough characterization of combustible fraction of legacy waste are available and the results of those studies are reported in Table 12.3. In the Indian context, two notable studies specifically focused on this area. The first study conducted by Cheela et al. (2021) analyzed the combustible fraction of legacy waste from an MSW landfill located in Visakhapatnam city (Andhra Pradesh). The samples were collected from different locations where the age of waste ranged from less than 5–15 years. Their findings indicated an increase in moisture content, a decline in volatile matter, and calorific value with respect to an increase in waste age. The second study by Mankhair and Chandel (2024) performed on a MSW dumpsite in Pune (Maharashtra) attempted to evaluate the changes in physicochemical characteristics of combustible fraction with increasing landfill depth. Samples were collected up to a depth of 12 m from an ongoing

Table 12.3 Characteristics of the combustible fraction of legacy waste

Parameters[a]	Mankhair and Chandel (2024)	Cheela et al. (2021)[b]	Hernández Parrodi et al. (2020)	Quaghebeur et al. (2013)[c]
Proximate analysis				
Volatile matter (%)	63.9	63.2	–	–
Ash (%)	27.4	25.1	38.3	29.0
Fixed carbon (%)	8.7	11.7	–	–
Ultimate analysis				
Carbon (%)	56.5	71.6	39.7	41.7
Hydrogen (%)	8.9	0.9	5.3	5.4
Nitrogen (%)	1.2	0.8	1.2	0.7
Sulfur (%)	0.02	1.4	0.6	0.3
Total organic carbon (%)	–	–	40.7	46.4
Calorific value (MJ/kg)	19.2	20.5	14.4	18.2
Chlorine (%)	0.9	–	1.2	4.2

[a]All values are on a dry basis
[b]Values for sampling location (S1)
[c]Calculated for sample 1 (1980–1985) based on the composition of combustible fraction

landfill mining project. The analysis revealed a decline in calorific value with increasing depth, while other parameters, such as proximate and ultimate analyses and chlorine content, exhibited no distinct trends. Based on these findings, both studies concluded that WtE is the most feasible valorization approach for the combustible fraction of legacy waste.

12.3.1 Proximate Analysis

The parameters obtained from the proximate analysis of waste play are crucial in the case when it is intended to be used as feedstock for WtE treatment processes. High moisture and ash content results in a low calorific value of waste, limiting its suitability as a feedstock for WtE treatment. The knowledge of volatile matter and fixed carbon content is essential for optimizing combustion processes to maximize energy recovery efficiency. Waste with a high fixed carbon content requires an extended retention time for complete combustion. Characterization results of the combustible fraction of legacy waste revealed that it is usually associated with a high ash content (in the range of 25.1–38.3%) as compared to fresh waste (Table 12.3). The high ash content can be attributed to the contamination of this fraction with impurities and the degraded organic matter that adheres to it (Singh and Chandel 2020). The fixed carbon content in this fraction ranged between 8.7%

and 11.7% which is higher compared to the normal MSW, this increase is due to the contamination of this fraction by organic impurities in the landfill (Cheela et al. 2021).

12.3.2 Ultimate Analysis

The ultimate analysis provides the elemental composition of waste, which is crucial for evaluating its energy recovery potential. This data is also used to predict emissions during thermal treatment processes and optimize the process efficiency (Bosmans et al. 2014). The carbon content in the combustible fraction of legacy waste is generally lower than that in fresh MSW, primarily due to the loss of carbon during anaerobic decomposition within landfills (Mankhair and Chandel 2024; Singh and Chandel 2020). In a study by Tao et al. (2024), the excavated waste had a lower carbon and sulfur content, whereas the hydrogen and oxygen contents increased significantly with increasing landfilling time, while nitrogen content showed less variation. In contrast to these results, Cheela et al. (2021) reported high sulfur levels which were attributed to the formation of sulfur oxides, hydrogen sulfide, and volatile organic compounds during anaerobic decomposition processes within the landfill. The sulfur content in the combustible fraction is a critical parameter for assessing its suitability for thermal treatment. In incineration processes, sulfur oxidizes to sulfur dioxide (SO_2), which can further form sulfuric acid (H_2SO_4), posing a risk of corrosion of the incinerator walls (Zhang et al. 2008). Given this consideration, the maximum permissible sulfur content for the combustible waste to be utilized as RDF is 1.5% (CPHEEO 2018). All the reviewed studies satisfied the limit for sulfur levels, confirming the feasibility of utilizing the combustible fraction of legacy waste both as RDF and as feedstock for WtE processes.

12.3.3 Calorific Value

The calorific value of the combustible fraction of legacy waste is one of the important parameters in the case of its valorization in WtE treatment technologies. Mankhair and Chandel (2024) reported a calorific value of 19 MJ/kg for this fraction, noting a decrease with respect to the depth of the landfill. Similarly, Cheela et al. (2021) documented calorific values ranging from 11 to 20 MJ/kg, observing a decrease with increasing age of the legacy waste. Rotheut and Quicker (2017) also reported calorific values for RDF derived from legacy waste, ranging between 9 and 26 MJ/kg. Although the calorific value of the combustible fraction of legacy waste is generally lower than that of fresh waste, they still satisfy the requirements to be used as a feedstock for WtE treatment (Cheela et al. 2021; Mankhair and Chandel 2024; Quaghebeur et al. 2013). The lower calorific values are due to the high levels of contamination which results in high ash content in this fraction. Furthermore, the calorific value of this fraction can be improved by adopting pre-treatment techniques like cleaning, drying, and shredding.

12.3.4 Chlorine and Heavy Metal Content

Determining chlorine content in the combustible fraction of legacy waste is essential, as elevated levels can result in harmful emissions, such as dioxins, and cause corrosion in WtE equipment. Based on the guidelines by the Central Public Health and Environmental Engineering Organisation (CPHEEO), Government of India, for using the combustible fraction as RDF, the permissible chlorine content is up to 0.5%, 0.75%, and 1% for RDF of grade I, II, and III, respectively (CPHEEO 2018). In the studies reviewed, the reported chlorine levels were within these standards, confirming compliance for utilization as RDF.

The analysis of heavy metal content in the combustible fraction of legacy waste is essential for its valorization as WtM. However, limited studies have reported the heavy metal content in this fraction. A study from Germany by Hernández Parrodi et al. (2020), assessed the heavy metal content in the combustible fraction of legacy. The findings indicated that all heavy metals (except Pb) in it were within the limits specified by the Recycling Building Materials Ordinance (German standards) and was found to be suitable for use in power plants.

12.4 Potential Valorization Options

Excavated legacy waste can be valorized through two primary pathways, namely WtM and WtE. These approaches provide strategies to recover materials and energy from legacy waste, playing a vital role in advancing toward sustainable waste management practices.

12.4.1 Waste-to-Material

The plastic, paper, and wood fractions are typically heterogeneous and characterized by high levels of contamination, which significantly limit their suitability for direct material recovery. Similarly, the quality of the textile waste recovered from landfills also makes it challenging to reuse or recycle effectively. This contamination is primarily due to the impurities such as dirt and organic residues adhering to it (Quaghebeur et al. 2013; Zhou et al. 2014). However, adopting a few pre-treatment techniques can mitigate the contamination levels and improve the quality of recovered material, facilitating its valorization as WtM. Adopting these techniques not only facilitates resource recovery but also aligns with the principles of circular economy.

12.4.1.1 Material Recovery from Excavated Plastic Waste

Plastic fraction of the legacy waste offers several potential pathways for valorization as WtM. It can serve as a raw material for producing renewable plastics or undergo chemical treatment to recover valuable products such as char, coil, carbon monoxide, hydrogen, and other ignitable gases (Kumar et al. 2011). Technologies like pyrolysis and gasification, are utilized for the conversion of waste plastics into useful products like char, oil, and gas. The oil obtained through such processes can be recycled as raw material in petrochemical plants, creating a circular pathway for material recovery (Al-Salem et al. 2009).

Mechanical recycling is another potential option for utilizing the plastic fraction of legacy waste. This process involves several pre-treatment steps, including sorting, cutting, shredding, milling, washing and drying, agglutination, extrusion, and pelletization to prepare the material for reuse (Al-Salem et al. 2009). Plastic fraction of legacy waste can be used as a binder replacing cement in the production of construction materials. The excavated plastic can be melted and mixed with filler material to manufacture paving blocks, tiles, and bricks. Additionally, the plastic waste can be shredded into fiber that can be directly incorporated into the concrete as filler material, hence reducing the dependency on natural aggregates (Mankhair et al. 2024) (Fig. 12.2). A study by Goli and Singh (2023), prepared composite materials using landfill-mined soil-like fraction and excavated plastic waste (as a binder). The study demonstrated that composite materials made from excavated plastic waste exhibit superior mechanical properties compared to composites using fresh plastics. Despite these opportunities, the mechanical recycling of excavated plastic waste

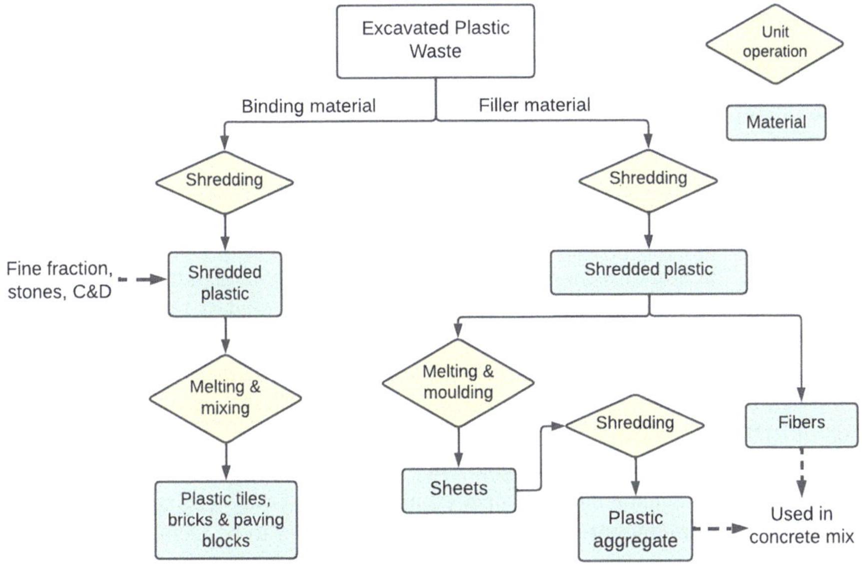

Fig. 12.2 Valorization pathways for plastic fraction of legacy waste

faces challenges, including the variability in its chemical and physical characteristics, presence of impurities, and marketability issues (Canopoli et al. 2018).

12.4.2 *Waste-to-Energy*

The combustible fraction of legacy waste possesses a high calorific value, making it suitable for transformation into RDF through pre-treatment like drying, shredding, and addition of a dechlorinating agent (Chen and Christensen 2010; Passamani et al. 2016; Tao et al. 2024). The RDF serves as an important intermediate product that can be utilized in thermal treatment processes with energy recovery (Danthurebandara et al. 2015). In a study by Mankhair and Chandel (2024), the combustible fraction of legacy waste was characterized by high ash content which limits its application in WtE treatment processes, and other parameters analyzed indicated its potential suitability for energy recovery through WtE treatment. The study suggested that pre-treatment techniques, such as shredding can be adopted to remove the adhered impurities. This will not only reduce the ash content but also increase the calorific value. Based on the characteristics of the combustible fraction, it is suitable for WtE treatment processes including incineration, pyrolysis, and gasification, as well as for use as RDF or co-processing with other waste materials in cement kilns.

12.4.2.1 Incineration

In India, for the incineration of combustible waste, the volatile matter should be greater than 40% while the fixed carbon and ash content and fixed carbon should be less than 35%, respectively (CPCB 2016). Additionally, waste with a net calorific value of more than 6.3 MJ/kg can potentially be used for WtE treatment techniques (MoEFCC 2016). In the majority of the referred studies, the combustible fraction meets these criteria, making it a feedstock for energy recovery (Cheela et al. 2021; Mankhair and Chandel 2024; Tao et al. 2024). A study by Mankhair and Chandel (2024) calculated the energy generation potential considering the incineration of the combustible fraction of legacy waste obtained from an Indian landfill. The findings of the study revealed an electrical energy generation potential of 1125 kWh per ton of combustible waste. Furthermore, the study also mentioned that by implementing a shredding pre-treatment for the removal of adhered impurities, the energy recovery potential could be enhanced. Rotheut and Quicker (2017) examined the utilization of RDF derived from the excavated legacy waste without any pre-treatment in a commercial-scale incineration plant. The study concluded that direct incineration of the RDF posed challenges in proper combustion, and only a mixture containing a sufficiently high amount of fresh waste was suggested to be technically feasible for incineration.

During incineration, flue gases such as CO_2, H_2O, O_2, and N_2 are produced. Depending on the composition of the incinerated waste and the operating conditions, smaller quantities of CO, SO_X, NO_X, HCl, HF, HBr, HI, VOCs, dioxins, and furans are released (Bosmans et al. 2013). Moreover, the combustible fraction of excavated legacy waste is associated with adhered impurities due to the contact with soil-like material and the generated leachate, resulting in elevated levels of heavy metals (Canopoli et al. 2018; Zhou et al. 2014). As a result, the use of RDF prepared from the combustible fraction of legacy waste needs proper consideration and measures to control the increased emissions along with effective pre-treatment. Rotheut and Quicker (2017) reported that conventional flue gas systems can handle these increased emissions; however, they emphasized the need for the effective performance of cleaning systems. The incineration of the combustible fraction of legacy waste results in a higher volume of residues, such as fly ash and bottom ash, compared to fresh MSW due to its elevated ash content. These residues are often concentrated with heavy metals; necessitating extra efforts for their safe management which makes the overall process expensive (Canopoli et al. 2018). However, waste incineration can be a suitable valorization approach if it is integrated with energy recovery, emission control, and proper disposal of residual waste.

12.4.2.2 Pyrolysis and Gasification

A study by Bosmans et al. (2013) suggested that the combustible fraction of legacy waste can be a suitable feedstock for utilization in pyrolysis and gasification systems. However, limited research studies exist on the pyrolysis and gasification of legacy waste and the characterization of the resulting products. Breyer et al. (2017) conducted co-pyrolysis of the excavated plastic waste with used lubrication oils to generate an alternative fuel for cement kilns. The highest performance was achieved with a 60% plastic content and reported that the process was economically viable in terms of energy efficiency. Zaini et al. (2019) studied the pyrolysis of RDF obtained from both excavated legacy waste and fresh MSW. The results from the pyrolysis tests indicated a lower yield of oil and gas (9–36% and 15–26%, respectively) from the legacy waste than that of fresh MSW samples (53% and 24%, respectively), primarily due to the high ash content.

Jagodzińska et al. (2021) conducted lab-scale pyrolysis of RDF (prepared from excavated waste) at six different temperatures (400, 450, 500, 550, 600, and 700 °C) and characterized the condensable and non-condensible pyrolytic products. The char yield decreased with increasing temperature, and the maximum yield was reported at 600 °C. Contrary to this, the change of condensibles and non-condensibles showed a non-linear variation with respect to temperature. The analysis revealed a high content of $C_2 - C_3$ hydrocarbons in the non-condensable fraction, while condensable products exhibited an abundance of polycyclic aromatic hydrocarbons. The elevated ash content in the excavated legacy waste reduces the overall yield of pyrolysis products, such as oil and gas, by promoting the formation of solid residues like char (Jagodzińska et al. 2021). Impurities, including heavy metals and non-organic materials, can deactivate catalysts, disrupt thermal decomposition pathways, and lead to the generation of contaminants in the pyrolysis products (Jagodzińska et al. 2022). These challenges necessitate pre-treatment methods to improve the quality of the feedstock, enhance process efficiency, and ensure the production of cleaner and more valuable pyrolysis outputs.

Chiemchaisri et al. (2010) tested the RDF prepared from the combustible fraction of legacy waste in a small-scale down-draft gasification system. The produced gas had a calorific value of 1.76 MJ/m^3 with a cold gas efficiency of 66%. In another study, Agon et al. (2016) conducted plasma gasification experiments on RDF obtained from excavated legacy waste in a pilot-scale reactor and compared the results with biomass gasification in the same system. The findings revealed that the syngas produced exhibited a calorific value of 10.9 MJ/Nm^3, minimal tar content, and high levels of CO and H_2. However, the gasification efficiency was lower compared to biomass, which was influenced by factors such as the ash and moisture content and the particle sizes of the RDF.

12.4.2.3 Co-processing in Cement Kilns

The RDF produced from the combustible fraction of the legacy waste can be used as a substitute for fossil fuel for co-combustion in cement kilns. The Indian Solid Waste Management (SWM) Rules (2016) recommends utilizing waste with a calorific value exceeding 12.5 MJ/kg for co-processing in cement kilns (CPHEEO 2018). Furthermore, based on the guidelines by CPHEEO, the waste with a calorific value greater than 18.8 MJ/kg is classified as RDF grade I, which is suitable for direct co-processing in cement kilns (Table 12.4). The calorific values reported in the studies by Cheela et al. (2021) and Mankhair and Chandel (2024) meet the criteria for classification as RDF grade I (Table 12.3). Numerous studies have demonstrated the feasibility of using RDF for co-combustion in cement kilns as an effective

Table 12.4 Requirements according to the guidelines for utilization of SCF and RDF in India (CPHEEO 2018)

Parameters	RDF Grade I	RDF Grade II	RDF Grade III	SCF*
Gross calorific value (KCal/kg)	>4500	>3750	>3000	>1500
Ash (%)	<10	<10	<15	<20
Moisture (%)	<10	<15	<20	<35
Chlorine (%)	<0.5	<0.7	<1.0	< 1.0
Sulfur (%)	<1.5			
Intended use	For direct co-processing in cement kiln		For co-processing with other waste materials in cement kiln	Feedstock for WtE plant

*Segregated Combustible Fractions

valorization pathway (Bhatnagar et al. 2017; Tao et al. 2024). In a study by Mankhair and Chandel (2024), based on the overall analyzed parameters of the combustible fraction, it was classified as RDF grade III, which can be used for co-processing with other waste materials in cement kilns. Despite this potential, the cement industries are reluctant to use the RDF originating from legacy waste due to quality concerns such as heterogeneity and contamination. To address this issue, a detailed characterization of the RDF and efforts to maintain consistent quality is essential, ensuring its acceptability and effective utilization.

12.5 SWOT Analysis of the Utilization of Combustible Fraction as Fuel

The studies on the characterization and valorization of combustible fraction of legacy waste highlight its potential for utilization as a fuel. However, this approach is associated with several benefits, limitations, and challenges. To evaluate its feasibility for utilization as a fuel, potential impacts, and challenges associated with it, a SWOT analysis was performed for the Indian context. It helps in identifying the strengths and opportunities that highlight the advantages of energy recovery and resource valorization, while weaknesses and threats ensure that challenges such as environmental and human health risks, economic barriers, and regulatory issues are effectively addressed in future studies. This holistic approach facilitates the development of sustainable strategies for the utilization of waste as a fuel, ensuring that resource recovery aligns with environmental and socio-economic goals. The current section presents an analysis of strengths, weaknesses, opportunities, and threats for the usage of combustible fraction of legacy waste as fuel.

12.5.1 Strengths

12.5.1.1 High Energy Potential

The combustible fraction has a high calorific value, making it suitable for use as feedstock energy recovery in thermal treatment processes. This makes it a valuable resource for heat and power generation, reducing the dependency on conventional fossil fuels.

12.5.1.2 Availability of RDF as a Feedstock

The combustible fraction is available in large quantities from the ongoing landfill mining projects, RDF prepared using this fraction offers a consistent feedstock for thermal treatment processes and co-processing in cement kilns.

12.5.1.3 Compliance and Policy Support

Government regulations like the Indian SWM Rules 2016 and initiatives like Swacch Bharat Mission promote the utilization of high calorific value waste in thermal treatment processes. Additionally, CPHEEO issued separate guidelines in 2018 for using RDF in various industries. These factors support the utilization of RDF, promote the development of WtE facilities, and ensure compliance with regulations.

12.5.2 Weaknesses

12.5.2.1 Composition of Legacy Waste

The composition of the legacy waste is heterogeneous, which significantly depends on the origin of waste and its age. This leads to challenges in process optimization and inconsistent energy output during thermal treatment. In addition, legacy waste is characterized by elevated moisture and ash content which, needs extensive sorting and pre-treatment for transforming into RDF.

12.5.2.2 Operational Challenges

The RDF offers numerous benefits, however, its production and utilization is associated with several challenges. Pre-processing requirements, such as sorting, drying, and quality control, can be resource-intensive. The presence of contaminants like chlorine, heavy metals, and other impurities in the legacy waste can lead to equipment corrosion, operational inefficiencies, and increased emissions during thermal treatment, necessitating advanced pollution control technologies.

12.5.2.3 Lack of Trust in the Produced Fuel

In case the produced fuel is to be used for co-processing in cement kilns, cement industries have a lack of trust in the quality of RDF. Since the cement industries prioritize the quality of cement production and are not primarily concerned with MSW management, they are hesitant to engage in RDF co-processing due to concerns about low RDF quality. The low quality of RDF could potentially impact their production processes or the quality of the final product (cement). Specifically, cement plants are more concerned about high chlorine content in RDF and inconsistent calorific values, leading them to use additional alternative fuels like biomass.

12.5.2.4 Lack of Financial Incentives

The absence of transitional financial support to bridge the gap between RDF production costs and acceptable sales prices to cement companies hinders the development of a self-sustaining RDF market. Reducing RDF processing and co-processing costs below coal costs is crucial to establish a viable market for RDF co-processing in cement kilns. To achieve this, substantial investments in technology and value chain development are necessary for RDF production on a large scale.

12.5.3 *Opportunities*

12.5.3.1 Substitution for Fossil Fuels

The combustible waste has the potential to be used as an alternative source of fuel, particularly in energy-intensive industries like cement kilns. This will reduce the dependency on traditional fossil fuels and align with the global efforts for transitioning to cleaner energy systems.

12.5.3.2 Industrial Co-processing

Co-processing in cement kilns and power plants provides an efficient way to utilize the combustible fraction while reducing the carbon footprint of industrial processes. It also offers industries a cost-effective alternative to conventional fuels.

12.5.3.3 Job Creation

Establishing RDF production facilities significantly contributes to job creation within the waste management sector. The process of manufacturing RDF involves several labor-intensive steps, which can generate employment opportunities at both skilled and unskilled levels. In addition, the RDF processing facilities require technical staffs for plant operation, maintenance, and quality control, which further boost the job creation.

12.5.4 *Threats*

12.5.4.1 Environmental Pollution and Human Health Risks

Thermal treatment of RDF can release harmful pollutants such as dioxins, furans, particulate matter, and heavy metals, raising concerns about the environment and public health. These emissions can be effectively managed through air pollution control technologies, which will increase the capital, operational, and maintenance costs. In the thermal processing of RDF, one of the primary environmental concerns is the potential formation and release of dioxins and furans (collectively referred to as dioxins). These highly toxic, persistent organic pollutants (POPs) emerge under certain conditions during thermal treatment, especially when RDF includes chlorine-rich materials such as plastics (e.g., PVC) (Baca et al. 2023). Dioxins are majorly formed by two pathways: precursor reactions at 300–600 °C, where chlorinated organics combine, and de novo synthesis at 200–400 °C, where chlorine, carbon, and oxygen react on catalytic surfaces like copper or iron (Zhao et al. 2022). When RDF is co-combusted in cement kilns, the elevated temperatures (1400–2000 °C) and extended residence times generally break down dioxin precursors, with the alkaline conditions created by limestone further curbing chlorine availability and thus limiting formation (Karstensen 2008). Similarly, modern incinerators operating above 850 °C with a minimum 2-s residence time (as per EU standards) decompose these compounds effectively (Neuwahl et al. 2019). However, if RDF has high PVC content or combustion is suboptimal (e.g., poor oxygen supply), dioxins can persist or reform in cooler zones like preheaters or dust collectors, necessitating stringent controls (McKay 2002). Many countries adopt an emission limit of 0.1 ng TEQ/m^3 for dioxins, reflecting a widely recognized standard (CPHEEO 2018; Richards and Agranovski 2017).

The environmental risks associated with dioxins arise from their potent toxicity, long-lasting presence, and tendency to bioaccumulate, threatening both human health and ecological systems even at trace levels (Taylor et al. 2024). Through atmospheric deposition, dioxins can enter the food chain, concentrating in the fatty tissues of animals and humans, where they are associated with serious health effects, including cancer, reproductive disorders, and damage to the immune system (Zhao et al. 2022). Effective control of these emissions relies on the adoption of the best available pollution control technologies.

12.5.4.2 Economic Viability

The costs involved in the pre-treatment of waste to produce RDF, implementing thermal treatment technologies, and installing emissions control systems can be substantial. These expenses impact the economic feasibility of WtE projects, potentially limiting their affordability and scalability, especially for smaller or less-funded initiatives.

12.5.4.3 Public Opposition

Although the RDF seems to be a prominent alternative to fossil fuels for energy generation, communities often oppose the establishment of thermal treatment facilities. This is due to perceived environmental and health risks, potentially delaying projects and increasing operational challenges.

12.6 Conclusions and Future Perspective

The physicochemical properties of the combustible fraction play a major role in exploring the potential valorization options. The combustible fraction of legacy waste is associated with high contamination levels, and hence, recycling is typically not recommended for this fraction. The combustible fraction of legacy waste is characterized by a lower calorific value and a high ash content as compared to the fresh MSW, it still satisfies the requirements to be used as a fuel. Therefore, given its physicochemical properties, its utilization as RDF in WtE treatment and for co-processing in cement kilns offers the most promising valorization option. However, additional pre-treatment (like sorting, cleaning, and shredding) methods may be used to enhance its quality.

According to the SWOT analysis, considering the strength and opportunities, the combustible fraction of legacy waste proves to be a good fuel, with the potential to replace fossil fuel for energy recovery and industrial co-processing. However, challenges such as maintaining RDF quality and ensuring effective emission control during thermal treatment must be addressed to enhance its adaptability for utilization as RDF. Future research studies should focus on evaluating alternate WtE treatment processes to determine the most sustainable approach for the valorization of combustible fraction, considering both the environmental and economic factors.

References

Agon N, Hrabovský M, Chumak O, Hlína M, Kopecký V, Mašláni A, Bosmans A, Helsen L, Skoblja S, Van Oost G, Vierendeels J (2016) Plasma gasification of refuse derived fuel in a single-stage system using different gasifying agents. Waste Manag 47:246–255. https://doi.org/10.1016/j.wasman.2015.07.014

Al-Salem SM, Lettieri P, Baeyens J (2009) Recycling and recovery routes of plastic solid waste (PSW): a review. Waste Manag 29:2625–2643. https://doi.org/10.1016/j.wasman.2009.06.004

Baca D, Monroy R, Castillo M, Elkhazraji A, Farooq A, Ahmad R (2023) Dioxins and plastic waste: a scientometric analysis and systematic literature review of the detection methods. Environ Adv 13:100439. https://doi.org/10.1016/j.envadv.2023.100439

Bhatnagar A, Kaczala F, Burlakovs J, Kriipsalu M, Hogland M, Hogland W (2017) Hunting for valuables from landfills and assessing their market opportunities a case study with Kudjape landfill in Estonia. Waste Manag Res J Sustain Circ Econ 35:627–635. https://doi.org/10.1177/0734242X17697816

Bosmans A, De Dobbelaere C, Helsen L (2014) Pyrolysis characteristics of excavated waste material processed into refuse derived fuel. Fuel 122:198–205. https://doi.org/10.1016/J.FUEL.2014.01.019

Bosmans A, Vanderreydt I, Geysen D, Helsen L (2013) The crucial role of waste-to-energy technologies in enhanced landfill mining: a technology review. J Clean Prod 55:10–23. https://doi.org/10.1016/J.JCLEPRO.2012.05.032

Breyer S, Mekhitarian L, Rimez B, Haut B (2017) Production of an alternative fuel by the co-pyrolysis of landfill recovered plastic wastes and used lubrication oils. Waste Manag 60:363–374. https://doi.org/10.1016/J.WASMAN.2016.12.011

Canopoli L, Fidalgo B, Coulon F, Wagland ST (2018) Physico-chemical properties of excavated plastic from landfill mining and current recycling routes. Waste Manag. https://doi.org/10.1016/j.wasman.2018.03.043

Cheela VRS, John M, Dubey B (2021) Quantitative determination of energy potential of refuse derived fuel from the waste recovered from Indian landfill. Sustain Environ Res 31:24. https://doi.org/10.1186/s42834-021-00097-5

Chen D, Christensen TH (2010) Life-cycle assessment (EASEWASTE) of two municipal solid waste incineration technologies in China. Waste Manag Res 28:508–519. https://doi.org/10.1177/0734242X10361761

Chiemchaisri C, Charnnok B, Visvanathan C (2010) Recovery of plastic wastes from dumpsite as refuse-derived fuel and its utilization in small gasification system. Bioresour Technol 101:1522–1527. https://doi.org/10.1016/j.biortech.2009.08.061

CPCB (2016) Selection criteria for waste processing technologies. Central Pollution Control Board. Ministry of Environment, Forest and Climate Change, Government of India, New Delhi, India. https://cpcb. nic.in/uploads/MSW/SW_treatment_Technologies.pdf. Accessed 3 Dec 2024

CPHEEO (2018) Guidelines on usage of refuse derived fuel in various industries. Central Public Health and Environmental Engineering Organisation, Ministry of Housing and Urban Affairs, Government of India, New Delhi, India. https://cpheeo.gov.in//upload/5bda791e5afb3SBMRDFBook.pdf. Accessed 3 Dec 2024

Danthurebandara M, Van Passel S, Machiels L, Van Acker K (2015) Valorization of thermal treatment residues in enhanced landfill mining: environmental and economic evaluation. J Clean Prod 99:275–285. https://doi.org/10.1016/J.JCLEPRO.2015.03.021

García López C, Ni A, Hernández Parrodi JC, Küppers B, Raulf K, Pretz T (2019) Characterization of landfill mining material after ballistic separation to evaluate material and energy recovery potential. Detritus 8:5–23. https://doi.org/10.31025/2611-4135/2019.13780

Goli VSNS, Singh DN (2023) Valorization of landfill mined plastic waste and soil-like fractions in polymer composites—a comprehensive solution for sustainable landfill mining. J Clean Prod 420:138349. https://doi.org/10.1016/J.JCLEPRO.2023.138349

Hernández Parrodi JC, Vollprecht D, Pomberger R (2020) Case study on enhanced landfill mining at Mont-Saint-Guibert landfill in Belgium: physico-chemical characterization and valorization potential of combustibles and inert fractions recovered from fine fractions. Detritus:44–61. https://doi.org/10.31025/2611-4135/2020.13941

Jagodzińska K, Zaini IN, Svanberg R, Yang W, Jönsson PG (2021) Pyrolysis of excavated waste from landfill mining: characterisation of the process products. J Clean Prod 279:123541. https://doi.org/10.1016/J.JCLEPRO.2020.123541

Jagodzińska K, Jönsson PG, Yang W (2022) Pyrolysis and in-line catalytic decomposition of excavated landfill waste to produce carbon nanotubes and hydrogen over Fe- and Ni-based catalysts—investigation of the catalyst type and process temperature. Chem Eng J 446:136808. https://doi.org/10.1016/j.cej.2022.136808

Jain M, Kumar A, Kumar A (2023) Landfill mining: a review on material recovery and its utilization challenges. Process Saf Environ Prot 169:948–958. https://doi.org/10.1016/j.psep.2022.11.049

Kaartinen T, Sormunen K, Rintala J (2013) Case study on sampling, processing and characterization of landfilled municipal solid waste in the view of landfill mining. J Clean Prod 55:56–66. https://doi.org/10.1016/J.JCLEPRO.2013.02.036

Karstensen KH (2008) Formation, release and control of dioxins in cement kilns. Chemosphere 70:543–560. https://doi.org/10.1016/j.chemosphere.2007.06.081

Kumar S, Panda AK, Singh RK (2011) A review on tertiary recycling of high-density polyethylene to fuel. Resour Conserv Recycl 55:893–910. https://doi.org/10.1016/J.RESCONREC.2011.05.005

Mankhair RV, Chandel MK (2024) Investigating the characteristics of combustible fraction of legacy waste: a study on energy recovery potential and GHG emission quantification. Environ Res 251:118669. https://doi.org/10.1016/J.ENVRES.2024.118669

Mankhair RV, Singh A, Chandel MK (2024) Characterization of excavated plastic waste from an Indian dumpsite: investigating extent of degradation and resource recovery potential. Waste Manag Res J Sustain Circ Econ. https://doi.org/10.1177/0734242X231219654

McKay G (2002) Dioxin characterisation, formation and minimisation during municipal solid waste (MSW) incineration: review. Chem Eng J 86:343–368. https://doi.org/10.1016/S1385-8947(01)00228-5

MoEFCC (2016) Solid waste management Rules, 2016. Central Pollution Control Board. Ministry of Environment, Forest and Climate Change, Government of India, New Delhi, India. https://cpcb.nic.in/rules-2/. Accessed 3 Dec 2024

Neuwahl F, Cusano G, Benavides JG, Holbrook S, Roudier S (2019) Best Available Techniques (BAT) reference document for waste incineration; EUR 29971 EN. https://doi.org/10.2760/761437

Passamani G, Ragazzi M, Torretta V (2016) Potential SRF generation from a closed landfill in northern Italy. Waste Manag 47:157–163. https://doi.org/10.1016/J.WASMAN.2015.07.024

Prechthai T, Padmasri M, Visvanathan C (2008) Quality assessment of mined MSW from an open dumpsite for recycling potential. Resour Conserv Recycl 53:70–78. https://doi.org/10.1016/j.resconrec.2008.09.002

Quaghebeur M, Laenen B, Geysen D, Nielsen P, Pontikes Y, Van Gerven T, Spooren J (2013) Characterization of landfilled materials: screening of the enhanced landfill mining potential. J Clean Prod 55:72–83. https://doi.org/10.1016/j.jclepro.2012.06.012

Richards G, Agranovski IE (2017) Dioxin-like pcb emissions from cement kilns during the use of alternative fuels. J Hazard Mater 323:698–709. https://doi.org/10.1016/j.jhazmat.2016.10.040

Rong L, Zhang C, Jin D, Dai Z (2017) Assessment of the potential utilization of municipal solid waste from a closed irregular landfill. J Clean Prod 142:413–419. https://doi.org/10.1016/j.jclepro.2015.10.050

Rotheut M, Quicker P (2017) Energetic utilisation of refuse derived fuels from landfill mining. Waste Manag 62:101–117. https://doi.org/10.1016/J.WASMAN.2017.02.002

Singh A, Chandel MK (2020) Effect of ageing on waste characteristics excavated from an Indian dumpsite and its potential valorisation. Process Saf Environ Prot 134:24–35. https://doi.org/10.1016/J.PSEP.2019.11.025

Singh A, Mankhair RV, Chandel MK (2022) Open dumps and circular economy: assessing landfill mining potential, benefits and challenges for Indian dumpsites. In: Circular economy in municipal solid waste landfilling: biomining & leachate treatment. Springer, pp 47–63. https://doi.org/10.1007/978-3-031-07785-2_3

Somani M, Datta M, Ramana G, Sreekrishnan T (2018) Investigations on fine fraction of aged municipal solid waste recovered through landfill mining: case study of three dumpsites from India. Waste Manag Res J Sustain Circ Econ 36:744–755. https://doi.org/10.1177/0734242X18782393

Tao J, Liu Y, Kumar A, Chen G, Sun Y, Li J, Guo W, Cheng Z, Yan B (2024) Effect of landfilling time on physico-chemical properties of combustible fractions in excavated waste. Sci Total Environ 917:170371. https://doi.org/10.1016/J.SCITOTENV.2024.170371

Taylor MD, Langdon KA, Smith JA, Stevenson G, Edge K (2024) Polychlorinated dibenzodioxins/furans and dioxin-like polychlorinated biphenyls in fish and crustaceans of a recreationally fished estuary, following targeted remediation. Sci Total Environ 921:171089. https://doi.org/10.1016/j.scitotenv.2024.171089

Zaini IN, García López C, Pretz T, Yang W, Jönsson PG (2019) Characterization of pyrolysis products of high-ash excavated-waste and its char gasification reactivity and kinetics under a steam atmosphere. Waste Manag 97:149–163. https://doi.org/10.1016/J.WASMAN.2019.08.001

Zhang Y, Chen Y, Meng A, Li Q, Cheng H (2008) Experimental and thermodynamic investigation on transfer of cadmium influenced by sulfur and chlorine during municipal solid waste (MSW) incineration. J Hazard Mater 153:309–319. https://doi.org/10.1016/J.JHAZMAT.2007.08.054

Zhao B, Hu X, Lu J (2022) Analysis and discussion on formation and control of dioxins generated from municipal solid waste incineration process. J Air Waste Manag Assoc 72:1063–1082. https://doi.org/10.1080/10962247.2022.2100843

Zhou C, Fang W, Xu W, Cao A, Wang R (2014) Characteristics and the recovery potential of plastic wastes obtained from landfill mining. J Clean Prod 80:80–86. https://doi.org/10.1016/J.JCLEPRO.2014.05.083

Chapter 13
Characterisation and Reuse Potential of Refused-Derived Fuel (RDF) Obtained from Landfill Mining

Ankur Saini

13.1 Introduction

13.1.1 Background on Landfill Mining and RDF Production

Landfill mining has emerged as a sustainable practice for recovering valuable resources from existing landfills, aligning with circular economy principles. This process involves the excavation and processing of previously landfilled waste to retrieve materials for reuse, recycling, or energy recovery. Among the materials recovered, refuse-derived fuel (RDF) has gained attention due to its potential as an alternative fuel source. RDF is produced by screening, classifying, shredding, and conditioning waste materials, such as non-recyclable plastics, textiles, and paper (Lechtenberg and Diller 2012).

The production of RDF from landfill mining is not only a means of reducing landfill volumes but also provides an avenue for waste valorisation. RDF serves as a key input for energy recovery processes, such as in cement kilns and waste-to-energy (WtE) plants, contributing to resource efficiency and sustainability goals. Compared to traditional energy sources like coal, RDF offers lower net greenhouse gas emissions due to its biogenic content. Additionally, RDF helps reduce dependency on virgin fossil fuels, aligning with circular economy principles while providing comparable calorific values for specific industrial processes (Lechtenberg and Diller 2012).

Despite these benefits, significant challenges remain. High costs of landfill mining, safety risks from methane release, and variability in RDF composition, especially from older landfills, need to be addressed. For instance, RDF from mined

A. Saini (✉)
MVW Lechtenberg & Partner, Duisburg, Germany
e-mail: ankur.saini@lechtenberg-partner.de

M. Somani et al. (eds.), *Landfill Mining of Legacy MSW: Global Insights and Indian Experience*, Springer Transactions in Civil and Environmental Engineering, https://doi.org/10.1007/978-981-95-3959-8_13

landfills often has lower proportions of decomposable organics like textiles and paper due to extensive decomposition over time.

Valorisation of RDF offers noteworthy environmental and economic advantages. From an environmental standpoint, it reduces greenhouse gas (GHG) emissions by displacing fossil fuels and curbing methane emissions from landfills. Economically, RDF provides cost savings via energy recovery and revenue generation as an alternative energy source.

Integrating RDF valorisation into sustainable WtE practices is essential for tackling global waste management challenges. With rapid urbanisation and increased waste generation, innovative solutions like RDF valorisation are increasingly critical. They not only enhance energy security but also support global decarbonisation efforts. Furthermore, RDF valorisation helps industries comply with stringent environmental regulations while reducing operational costs by using locally available waste resources.

13.1.2 Scope and Objectives of This Chapter

This chapter outlines the processes involved in landfill mining for RDF production, the characterisation of RDF, focusing on its environmental and economic impact in the Indian cement industry. It also discusses challenges in RDF valorisation and potential pathways for improvement in the context of substantially reducing fossil fuel consumption.

13.2 Challenges in Landfill Mining

13.2.1 Methane, Gas Risks, and Landfill Instability

Excavating old landfills releases trapped methane and other landfill gases, which can reach life-threatening or explosive concentrations. Mitigation measures include:

- Controlled aeration during excavation.
- Continuous gas monitoring with alarms for methane concentrations.
- Proper ventilation systems to dissipate gases.

These risks highlight the need for strict adherence to health and safety (H&S) protocols.

Excavating legacy waste requires proactive methane management:

- Continuous monitoring with gas detection systems.
- Installation of passive and active ventilation systems.
- Controlled aeration to minimise explosive risks.

The risk of landfill slides or settlement poses also a significant challenge. For example, the Koshe landfill disaster in Addis Ababa in 2017, which killed 116 people, underscores the importance of geotechnical stability assessments prior to excavation (Vollmann 2019). Precautions such as slope stability monitoring and phased excavation are essential to prevent such tragedies.

Old landfills may contain hazardous materials, such as hospital waste, explosives, or chemical contaminants. Lessons from the Beirut project (2004–2006) emphasise the importance of pre-mining surveys to identify and manage such risks (Lechtenberg 2007). Specific precautions include:

- Use of personal protective equipment (PPE).
- Segregation and safe disposal of hazardous materials.
- Collaboration with specialists in hazardous waste management.

13.2.2 Economic Challenges: High Costs

The costs associated with landfill mining are significantly higher than those of standard RDF production from fresh waste. Excavation and processing in old landfills introduce challenges, including:

- High wear and tear on equipment: Foreign materials such as stones, glass, sand, and other heavy residues accelerate tear especially in shredders and screening machines.
- Contamination: Sand, small plastic particles, and other residues often remain in the excavated waste, requiring additional processing.

13.3 Overview of Landfill Mining for RDF Production

13.3.1 Landfill Mining Processes

Landfill mining employs techniques such as bio-remediation and bio-mining to excavate and process landfill waste. These methods stabilise, segregate, and recover resources from the landfill or old dumps.

The "Guidelines for Disposal of Legacy Waste (Old Municipal Solid Waste)" (CPCB 2019) recommended preliminary studies in order to ensure the landfill mining project is tailored to the specific characteristics of the site. For instance, understanding the landfill topography, environmental conditions, and waste composition helps optimise the techniques and technologies employed, ultimately improving the project's efficiency and outcomes. Before initiating bio-mining, a thorough precursor study is required. This includes:

- Site survey: Using drone mapping or total station surveys to analyse the landfill topography.
- Baseline environmental study: Evaluating parameters such as soil and leachate quality, trace elements concentrations, chlorides, sulphates, acidity, rainfall, surface hydrology, and wind direction.

- Waste characterisation: Understanding the composition and age of the waste to optimise bio-mining techniques.

13.3.1.1 Bio-remediation Process

1. Excavation and aeration:
    - Legacy waste is excavated and formed into windrows approximately 2 m high to maximise exposure to air.
    - Methane and other gases trapped in the waste are released, while leachate is dried through solar exposure.
    - Bio-cultures are mist-sprayed onto the waste to accelerate microbial decomposition and reduce odours.
2. Stabilisation:
    - Decomposition within the windrows generates biological heat, drying the waste through exposing through sun and reducing its volume.
    - Waste is considered stabilised when there is no heat, gas, or leachate generation, and seeds can germinate in the material.

13.3.1.2 Bio-mining Process

After bio-remediation bio-mining is initiated. This method includes the following:

1. Excavation of the dumped material by means of an excavator (as shown in the next figure.
2. Screening: Landfill waste is fed into a mobile screen, separating the material into two fractions or three fractions depending on the mobile screen type. As not all the materials could be utilised for RDF production, it is subject to mobile screening, where three fractions are obtained (also see next figure):
    - Fraction 1—minus 30 mm: Soil and inert fractions are moved for filling or to construction sites. These fractions include broken glass, stones, and rubble.
    - Fraction 2—(30–80) mm and Fraction 3—(80+) mm: Contains RDF fraction along with recyclables like metal pieces and non-combustible materials like inert, that could be transported to the RDF processing facility for further processing.

The operational costs for the Kleemann screening machine from Wirten group are as follows (Mohanty 2024, Enquiry of Kleemann mobile screen from Wirtgen India Pvt. Ltd. Personal interview):

- Maintenance costs: The cost of wear and spear parts is around 900 INR/h.
- Diesel powered machine consumes 12–14 L per hour at a diesel price of 88 INR/litre, the cost is in range of 1000–1250 INR/h.
- The energy cost of an electric-driven machine is around 500 INR/h (Fig. 13.1).

Fig. 13.1 Landfill waste being screened in three different fractions by Kleemann in Okhla landfill in Delhi, India

Fig. 13.2 A front loader feeds MSW into a mobile screen (manufacturer: German-based Doppstadt). The conveyor on the left-hand side shows the outlet of the screen passing, which is mainly organic matter (photograph on the right)

Mobile trommel or vibrating screen are preferred at landfill site for their flexibility and efficiency to relocate across the landfill, reducing the need to transferred unscreened material back to the landfill. As per the author's experience it recommended to place a mobile drum screen in the landfill directly and only transfer the overflow to the dedicated RDF production facility. While mobile screens are more versatile and reduce transport costs, stationary screens offer higher throughput and better sorting precision for large-scale operations. Mobile screens are typically powered by integrated diesel engines but can also operate on electricity if available, making them suitable for dynamic landfill environments. This setup minimises transport requirements, streamlines operations, and enhances the overall efficiency of RDF production, particularly in smaller or less consistent waste streams.

The following pictures show an example of such an installation MVW Lechtenberg & Partner implemented a few years ago in a landfill in Pakistan (Fig. 13.2).

After the waste has been screened at the landfill site, still it contains heavyweight matter, sand, some metal pieces and other inert fractions that need to be screened and separated from the combustibles.

13.3.2 RDF Production Facility

The "pre-screened MSW" refers to municipal solid waste that has already been subjected to a mechanical screening process at the landfill site as shown in Fig. 13.3 before being transferred to the stationary RDF production plant. This pre-screening is typically carried out using mobile trommel or vibrating screens that divide the waste into two primary fractions based on particle size. The fine fraction (<80 mm), consisting mainly of inert materials such as sand, soil-like matter, and broken glass, is removed and excluded from further RDF processing. The screened waste or coarse fraction from the landfill site could be further processed in a stationary RDF production plant for separating the combustibles. A typical layout of a stationary RDF production plant is shown in Fig. 13.4.

In a typical RDF production plant as shown in the next figure, a front loader feeds the aged and pre-screened waste into the following:

- *Primary shredder*: for opening the bags or cutting the waste into defined sizes of around 250–300 mm to facilitate subsequent sorting and separation processes.
- *Drum screen or splitter screen*: to separate the remaining inert fraction size of 80 mm from the landfill waste (Fig. 13.5).

- *Metal separator*: the remaining material after passes a ferrous separator before it enters the air classifier.
- *Air classifier*: This machine removes the hard/heavy particles from the light fraction. At this stage, hard materials like stones, broken glass, ceramics or similar might be present in the incoming waste stream and must be diverted from the combustibles. The light fraction is the desired RDF which is then conveyed to the

Fig. 13.3 A view of the outlet of the overs of the drum screen, which is mainly packaging materials, but also dried bushes and twigs (left). The right photograph shows a modification, where a manual sorter has been placed there to separate some recyclable matter. (Source: MVW Lechtenberg & Partner)

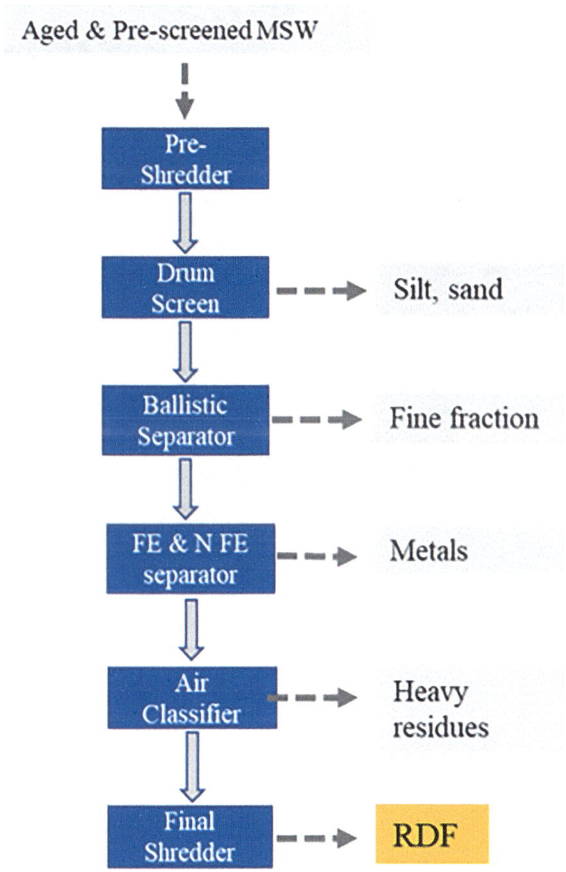

Fig. 13.4 A typical RDF process flow for aged and pre-screened MSW

final shredder. Moreover, the air classifier prevents entering any material into the final shredder, which might be deleterious for the comminution tools.

- *Ballistic separator*: In the ballistic separator, the main material is divided into three categories: 2D-fraction (lightweight fraction), 3D-fraction (heavyweight fraction), and fine material (<30 mm). The 2D and 3D materials could be utilised as RDF fractions depending on the composition of waste.
- *Shredder*: The shredder crushes the light fraction as well as the textiles down to the desired particle size that matches the requirements of the feeding point of the cement plant. It is assumed that particle sizes of roughly max. 70–80 mm are desirable for calciner firing in the cement plant.
- *Baler*: Shredded RDF could be transferred through a conveyor to a baler machine to press the material and prepare the bales to be transferred later to the cement plant. Baler is not necessary in every case depending on the logistic point of view.

The next figures depict a typical layout of an RDF production plant (Fig. 13.6).

Fig. 13.5 Left—Front end loader feeding the incoming material into a primary shredder; Right—Trommel screen in action to remove organic, slit, and other inert materials. (Source: MVW Lechtenberg & Partner)

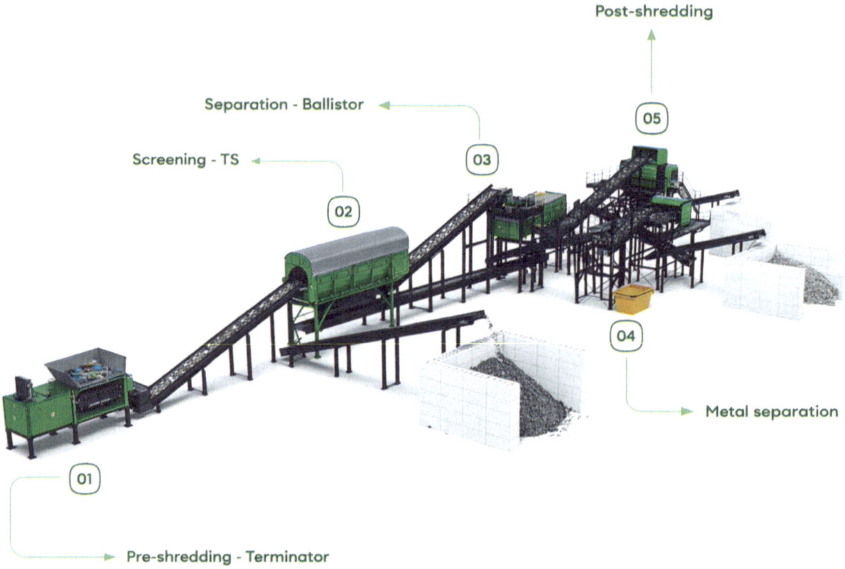

Fig. 13.6 A typical RDF production layout from Austrian-based company Komptech

13.4 Case Study: RDF Production in Lebanon from Landfill Waste

During the Lebanese civil war, significant amounts of waste were indiscriminately disposed of at the Beirut harbour site. In the aftermath, Solidere, a Lebanese company focused on developing Beirut's central district, took charge of clearing this waste to facilitate the city's reconstruction. Collaborating with MVW Lechtenberg & Partner, Solidere aimed to integrate environmentally friendly practices in waste management, focusing on the production RDF from landfill waste at Normandy landfill (Lechtenberg 2007) (Fig. 13.7).

Fig. 13.7 Vibrating screen in the Normandy landfill in Beirut. (Source: MVW Lechtenberg & Partner)

The project involved processing pre-sorted waste products with high calorific value into secondary fuels. The mixed waste included household textiles, foam, plastics, used tires, and wood. The implementation steps included:

- Mechanical and manual sorting: Waste was screened to separate inert materials such as sand for reuse in road construction.
- Metal separation: Magnetic separators were used to extract ferrous metals for recycling.
- Shredding: Combustible waste was reduced to smaller sizes using shredders. Pieces less than 30 mm were designated for main burner fuel, while those under 70 mm were allocated for pre-calciner fuel.
- Exporting RDF: As Lebanese cement plants lacked permits to use secondary fuels, the processed RDF was exported to neighbouring countries under Basel Convention regulations.

The project minimised landfill dependency, reduced methane emissions, and repurposed large volumes of waste for energy recovery. By converting waste into valuable RDF, Solidere achieved cost savings and reduced the environmental footprint of the reconstruction activities (Fig. 13.8).

Approximately 120,000 m^3 of waste was processed, clearing 60 ha of land for urban development. The project demonstrated the potential of RDF in achieving sustainable waste management and highlighted the importance of international cooperation for large-scale RDF valorisation efforts.

Fig. 13.8 Landfill waste being processed in Normandy landfill in Beirut. (Source: MVW Lechtenberg & Partner)

13.5 Characterisation of RDF from Landfill Mining

13.5.1 Legacy Waste Characterisation and Theoretical Yield of RDF

Legacy waste composition varies significantly based on landfill age and site-specific conditions. Components, such as plastics, textiles, and wood though degraded, still possess calorific value for RDF production. Unlike RDF produced from fresh waste, which typically includes non-recyclable plastics, textiles, and paper, the composition of RDF from old landfills is less consistent due to decomposition over time. Key points include:

- Reduced calorific value due to lower proportions of organics such as textiles, paper, and wood.
- Higher levels of inert materials like sand and rubble, which require additional separation.

The table below summarises the composition of legacy waste from four Indian landfills: Delhi, Kadapa, Pune, and Boragaon. The average values highlight the theoretical potential for RDF recovery (Table 13.1).

Plastics contribute the highest combustible share at 8.74%, followed by wood (1.18%), paper/cardboard (1.50%), textiles/rags (0.72%), and rubber/leather (0.50%). These fractions form the primary contributors to RDF due to their relatively high energy content.

Table 13.1 Average composition of aged MSW at selected Indian landfills

Waste fractions	Delhi landfill[a]	Kadapa landfill[a]	Pune landfill[b]	Boragaon landfill[c]	% Average aged MSW
Plastics	3.30%	3.70%	5.00%	22.95%	8.74%
Rubber / leather	0.00%	0.00%	1.00%	1.00%	0.50%
Paper / cardboard	0.00%	0.00%	6.00%	0.00%	1.50%
Textile / rags	0.80%	1.01%	0.00%	1.05%	0.72%
Wood	0.20%	1.30%	0.00%	3.20%	1.18%
Glass	0.20%	1.73%	5.00%	5.20%	3.03%
Metals	0.00%	0.00%	3.00%	0.85%	0.96%
Soil like material	71.90%	75.20%	33.00%	53.70%	58.45%
Others (non-combustibles)	0.20%	0.86%	22.00%	0.00%	5.77%
Construction & Demolition	23.40%	16.20%	25.00%	12.05%	19.16%
Sum	100.00%	100.00%	100.00%	100.00%	**100.00%**

Source: *Average aged landfilled composition data retrieved from the following
[a]Somani et al. (2018)
[b]Bhandare et al. (2024)
[c]Ghosh and Kartha (2023)

Table 13.2 Characterization of aged MSW in India

Waste fractions	% Average aged MSW[a]	% Sorting efficiency of recyclables[b]	% Recyclables could be extracted	% Sorting efficiency of machine for RDF[b]	Thereof, fractions useable as RDF
Plastics	8.74%	–	–	90.00%	7.86%
Rubber / leather	0.50%	–	–	90.00%	0.45%
Paper / cardboard	1.50%	–	–	90.00%	1.35%
Textile / rags	0.72%	–	–	90.00%	0.64%
Wood	1.18%	–	–	90.00%	1.06%
Glass	3.03%	–	–	–	
Metals	0.96%	90.00%	0.87%	–	
Soil like material	58.45%	–	–	–	
Others (non-combustibles)	5.77%	–	–	–	
Construction & demolition	19.16%	–	–	–	
Sum	**100.0%**		**0.87%**		**11.36%**

Source: [a]Average aged landfilled composition data retrieved from the following
1. Somani et al. (2018)
2. Ghosh and Kartha (2023)
3. Bhandare et al. (2024)
Note: [b]Sorting efficiency may vary based on equipment and site conditions. This data highlights the importance of optimising sorting technologies to maximise RDF and recyclables recovery

Conversely, non-combustible fractions such as soil-like material (58.45%), C&D waste (19.16%), glass (3.03%), metals (0.96%), and other inerts (5.77%) are not suitable for RDF production.

Based on the average composition, approximately 12.63% of the excavated waste has the potential to be converted into RDF, primarily from the combustible fractions mentioned above.

The following table provides an estimation of the useable fraction of RDF is 11.36% and recyclables is 0.87% at 90% sorting efficiency of screening and separation to recover maximum amount of combustible material (Table 13.2).

13.5.2 RDF Quality Parameters

RDF derived from landfill mining exhibits specific physical and chemical properties that determine its suitability as a fuel for cement plants. Key physical properties include (Lechtenberg and Diller 2012):

- Material size.
- Calorific value.
- Moisture and ash content.
- Chlorine and sulphur content.
- Trace elements.

In India, the regulations for the usage of RDF are primarily governed by the Ministry of Environment, Forest, and Climate Change (MoEFCC) and the Central Pollution Control Board (CPCB). In 2018, the Ministry of Housing and Urban Affairs (MoHUA) released guidelines on "Usage of Refuse Derived Fuel in Various Industries" under the Swachh Bharat mission (CPHEEO 2018). The recommendations present the following RDF standards for use in waste-to-energy plants and the cement industry, as confirmed by the Cement Manufacturing Association in India (Table 13.3).

Table 13.3 Indian RDF standards set by the Ministry of Housing and Urban Affairs in 2018 (CPHEEO 2018)

Parameters	SCF	RDF—Grade III	RDF—Grade II	RDF—Grade I
Intended use	Input material for the waste to energy plant or RDF preprocessing facility	For co-processing directly or after processing with other waste materials in cement kilns	For direct coprocessing in cement kilns	For direct co-processing in cement kilns
Size	Anything above 400 mm has to be mutually agreed between urban local body/SCF supplier and cement Plants.	<50 mm or <20 mm depending upon use in in-line calciner or separate line calciner, respectively		
Ash—maximum permissible	<20%[a]	<15%	<10%	<10%
Moisture—maximum permissible	<35%	<20%	<15%	<10%
Chlorine—maximum permissible	<1.0%[a]	<1.0%	<0.7%	<0.5%
Sulphur—maximum permissible	<1.5%[a]	<1.5%		
Net Calorific Value (NCV)[a] –(average value of every individual consignment)	>1500 kcal/kg net	>3000 kcal/kg net	>3750 kcal/kg net	> 4500 kcal/kg net
Any other parameter	SCF—any offensive odour to be controlled[b]	RDF—any offensive odour to be controlled	RDF—any offensive odour to be controlled	RDF—any offensive odour to be controlled

Note: [a]If the blending process is done in cement plants, the deviations in recommitted limit for ash, chlorine and sulphur content can be mutually agreed between urban local body/SCF supplier and cement plants

[b] Band width of variations acceptable in NCV can be mutually decided between the RDF manufacturer and cement plants

[c] Since odour is still largely a matter of perception and there is no satisfactory equipment to measure different types of odour, no quantitative figure has been given

MoHUA has established several ideal qualities for RDF for co-processing in cement plants and the production process. These are presented in the Central Public Health and Environmental Engineering Organisation (CPHEEO) Municipal Solid Waste Management Manual, PART II. As illustrated in the figure and points below, the parameters include:

- Chemical composition: RDF must meet certain standards for chemical composition, such as the ash content and calorific value, to ensure its suitability as a fuel.
- Impurities: RDF must meet standards for the presence of impurities, such as heavy metals, to prevent the release of toxic substances into the environment.
- Moisture content: RDF must meet standards for moisture content to ensure its stability and combustion efficiency as a fuel.
- Size and shape: RDF must meet standards for size and shape to ensure its suitability for handling, storage, and use as fuel (Fig. 13.9).

In Europe, EN ISO 21640:2021-11 is the latest norm used for the solid recovered fuels (SRF) specification and classes. It supersedes the former EN 15359:2012-01. It has been prepared by Technical Committee ISO/TC 300 "Solid recovered materials, including SRFs" in collaboration with Technical Committee CEN/TC 343 "Solid Recovered Fuels."

The CEN European standard for SRF is briefly introduced in the section that follows. The essential SRF characteristics are then discussed, from the viewpoint of particular end-users, in relation to how the mechanical biological treatment (MBT) processes should be designed and run. The large ongoing research and development effort for a European SRF quality assurance or control system includes the CEN technical standard for SRF specification and classes. The creation of this standard has been modified to satisfy the technical and legal needs of the customer, such as

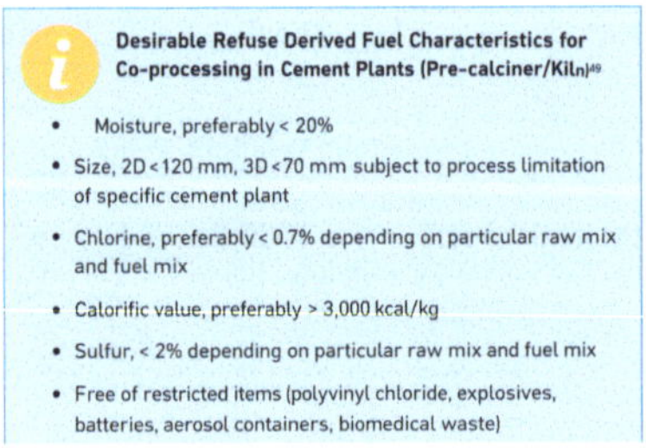
Desirable Refuse Derived Fuel Characteristics for Co-processing in Cement Plants (Pre-calciner/Kiln)[49]

- Moisture, preferably < 20%
- Size, 2D < 120 mm, 3D < 70 mm subject to process limitation of specific cement plant
- Chlorine, preferably < 0.7% depending on particular raw mix and fuel mix
- Calorific value, preferably > 3,000 kcal/kg
- Sulfur, < 2% depending on particular raw mix and fuel mix
- Free of restricted items (polyvinyl chloride, explosives, batteries, aerosol containers, biomedical waste)

Fig. 13.9 Parameters recommended by MoHUA for co-processing of RDF in cement plants in India (CPHEEO 2018)

adhering to the Waste Incineration Directive (WID) emission restrictions (EIB 2024). The achievable quality of SRF/RDF has also been taken into account. Instead of being input-oriented, it applies at the point where the SRF producer and intended end-user meet.

In the EU, the SRF is defined as follows (ISO 21640 (2021)):

- *SRF are produced from non-hazardous waste. The input waste can be production specific waste, municipal solid waste, industrial waste, commercial waste, construction and demolition waste, sewage sludge etc. It is thus obvious that SRF are a heterogeneous group of fuels. A well-defined system for classification and specification is therefore of great importance to reach the above mentioned objectives and intentions.*

In addition to that, SRF are categorised into classes focusing on the key properties that are defined by boundary conditions:

- Net calorific value
- Chlorine content
- Mercury

The classification system is based on three crucial factors, known as the primary SRF characteristics: a technical feature (chlorine content), an economic characteristic (net calorific value), and an environmental characteristic (mercury content). The traits are selected to provide an immediate but simplified picture of the fuel in question (Table 13.4).

As the table shows, the ISO 21640 defines a very broad range of SRF, which reflects a high caloric material with fewer pollutants on one end, and a material of very low calorific value with high pollutants on the other. Classes 4 and 5 would hardly be suitable in a clinker kiln (EIB 2024). Moreover, a quality-certified SRF does not necessarily imply a high fuel quality. Instead, it relates to a more consistent, continuously produced fuel that meets the quality demanded by end-users and their regulators.

Table 13.4 EU Classification for solid recovered fuels (ISO 21640 2021)

Classification characteristic	Statistical measure	Unit	Classes 1	2	3	4	5
Net calorific value (NCV)	Mean	MJ/kg (ar)	≥25.0	≥20.0	≥15.0	≥10.0	≥3.0
Chlorine (Cl)	Mean	% in mass (dm)	≤0.2	≤0.6	≤1.0	≤1.5	≤3.0
Mercury (Hg)	Median	mg/MJ (ar)	≤0.02	≤0.03	≤0.05	≤0.10	≤0.15
	80th percentile	mg/MJ (ar)	≤004	≤0.06	≤0.10	≤0.20	≤0.30

Note: ar = as received; dm = dry matter

13.6 RDF as a Fuel Source for Cement Plants in India

13.6.1 Potential of RDF Production and Usage in the Indian Cement Industry

India generated 1,70,339 tonnes per day of MSW in 2022, from that 41,455 tonnes per day were landfilled (CPCB 2022). This significant waste volume offers an untapped opportunity for energy recovery through the production of RDF. According to MSW generation data (CPCB 2022) and population projections (Statista 2025), per capita MSW generation is expected to rise further by 2029, highlighting the growing need for sustainable waste management solutions.

From the total MSW landfilled in 2022, approximately 1.72 million tonnes of waste (11.36% of MSW landfilled) can theoretically be converted into RDF annually as estimated in the section 5.1. However, this potential depends heavily on the efficiency of sorting and processing equipment.

India's cement industry is energy-intensive, with a substantial portion of its energy requirement met through coal. In fiscal year (FY) 2023, the total clinker production in India reached approximately 288.4 million tonnes (IBEF 2025), with an average clinker-to-cement ratio of 0.77 (ZKG 2015), resulting in total cement production of 374.55 million tonnes. Cement kilns require significant thermal energy, with an average heat consumption of 720 kcal per kilogramme of clinker (as per author experience from projects in India). Given the high thermal demand of cement kilns in India, predominantly fulfilled through coal, this energy demand translates to approximately 207.65 million Gcal annually. RDF presents an opportunity to replace a portion of this coal, contributing to both energy efficiency and emission reduction (Saha and Karstensen 2017).

The next table presents a comparison between the baseline scenario (100% coal) and a substitution scenario where 30% of thermal energy is sourced from RDF (Table 13.5). Key points include:

- Net calorific value (NCV):
  - Coal has a higher NCV of 6500 kcal/kg, while RDF's NCV is lower at 3000 kcal/kg due to its composition.
  - This means a higher quantity of RDF is needed to produce the same energy as coal.
- Dosage of fuels:
  - In the baseline scenario, 31.95 million tonnes of coal are required to meet the annual heat demand of cement kilns.
  - With 30% RDF substitution, 28.32 million tonnes of RDF and only 22.36 million tonnes of coal are required.
- Heat portions:
  - The baseline scenario utilises 207.65 million Gcal entirely from coal.
  - In the substitution scenario, RDF contributes 62.30 million Gcal, while coal provides the remaining 145.36 million Gcal.

Table 13.5 Thermal substitution potential to offtake RDF in Indian cement industry

Total clinker production in India FY23	[t/a]	288,403,500		
Specific heat consumption	[kcal/kg clinker]	720[a]		
Heat consumption	[Gcal/a]	207,650,520	Clinker production in FY23	288.4 million tonnes
Thermal substitution		**Baseline**	**Substitution scenario**	
		Coal	**RDF**	**Coal**
Net calorific value	[kcal/kg]	6500[b]	3000[c]	6500
Dosage fuels	[t/a]	31,946,234	20,765,052	22,362,364
Heat portions	[Gcal/a]	207,650,520	62,295,156	145,355,364
Substitution rate	[%]		30.0%	

Source
[a]Specific heat consumption of clinker data is retrieved from MVW Lechtenberg QIS database
[b]Net calorific value of coal data is retrieved from MVW Lechtenberg QIS database
[c]Net calorific value of RDF data is retrieved from MVW Lechtenberg QIS database

- Thermal substitution rate (TSR):
  - A 30% TSR reflects the proportion of energy derived from RDF, highlighting its potential to reduce reliance on coal.

Despite the high demand for RDF in the cement industry, approximately 20.76 million tonnes annually (calorific value of 3000 kcal/kg) at a 30% TSR. However, the current RDF extraction potential from landfilled MSW is limited to 1.72 million tonnes per year. This results in significant supply gap of approximately 19.05 million tonnes of RDF annually. However, the abundance of biomass, industrial plastic waste availability in India could fulfil this gap, mostly in the states of Maharashtra, Tamil Nadu, Uttar Pradesh, Gujarat, Andhra Pradesh, Karnataka, Rajasthan and Madhya Pradesh (Saha and Karstensen 2017). The necessity for improved waste segregation, collection, and processing infrastructure to bridge this gap requires investment.

13.6.2 Economic and Environmental Benefits of RDF Valorisation

The valorisation of RDF as an alternative fuel in cement plants offers a multitude of environmental benefits, addressing critical challenges such as greenhouse gas emissions, landfill dependency, and sustainable waste management.

Table 13.6 Thermal substitution estimation for RDF replacing coal in Indian cement industry

Coal			**RDF**		
NCV	6500	kcal/kg	NCV	3000	kcal/kg
Emission factor	96[a]	t CO_2 / TJ	Emission factor	60[b]	t CO_2 / TJ
(default IPCC)	2.61	t CO_2 / t coal	(around 40% biogenic share)	0.75	t CO_2 / t RDF
	Scenarios	**0**	**I**		
		Baseline: Coal	**30% substitution**		
Coal	[t/a]	31,946,234	22,362,364		
Emission CO_2 from coal	[tCO_2/a]	83,461,555	58,423,088		
RDF	[t/a]	0	28,315,980		
Emission CO_2 from RDF	[tCO_2/a]		15,649,042		
Total CO_2 emissions	[tCO_2/a]	83,461,555	74,072,130		
Difference to baseline	[tCO_2/a]		9,389,425		
	[%]		11.3%		

Source
[a] (IPCC 2006)
[b] Emission factor of RDF is retrieved from MVW Lechtenberg QIS database

13.6.2.1 Reduction in Greenhouse Gas (GHG) Emissions

Replacing coal with RDF significantly reduces fossil-derived CO_2 emissions due to its lower emission factor. Being entirely fossil, coal has an emission factor of 96 t CO_2/TJ (IPCC 2006), thus generating 2.61 t CO_2/tonne. On the other hand, RDF has a lower emission factor of 60 t CO_2/TJ, generating only 0.63 t of fossil CO_2 per tonne, accounting for its 40% biogenic share (for instance paper, cardboard, textiles made from natural fibres, wood) as per author's experience.

At a 30% (TSR) in cement plants of India, the annual CO_2 emissions can be reduced by approximately 9.38 million tonnes, representing an 11.3% decrease compared to the baseline scenario of coal-only usage. This reduction plays a crucial role in helping India meet its climate targets under the Paris Agreement by decreasing the carbon footprint of energy-intensive industries (Table 13.6).

13.6.2.2 Minimised Landfill Dependency

RDF production directly supports the diversion of significant fractions of MSW away from landfills, mitigating a range of environmental hazards:

- Groundwater contamination: Reduced landfill dependency minimises the leachate generation, which otherwise pollutes groundwater resources.

- Methane emissions: Landfills are a significant source of methane, a potent greenhouse gas. By extracting waste for RDF production, methane emissions can be effectively curtailed.
- Land use: Diverting waste fractions to RDF production reduces the need for new landfill sites, preserving valuable land resources and reducing the socio-environmental conflicts associated with landfill siting.

13.6.2.3 Sustainable Waste Management

Promoting RDF production aligns seamlessly with India's circular economy goals and the Swachh Bharat Mission by transforming waste into a valuable resource. Key benefits include:

- Resource recovery: Non-recyclable waste fractions, such as plastics, textiles, and rubber, are converted into RDF, contributing to waste valorisation and resource efficiency.
- Reduction in open burning: Diverting waste for RDF production reduces incidents of open waste burning, which significantly contributes to air pollution and public health hazards.

13.6.2.4 Mitigating Climate Change

By partially substituting coal with RDF in cement plants, the reliance on high-carbon fossil fuels is reduced. This transition supports the global shift towards cleaner energy alternatives, aiding in the reduction of climate change impacts.

Despite the promising benefits, certain challenges must be addressed to fully realise the environmental potential of RDF:

- Improving waste segregation efficiency: The efficiency of waste segregation technologies remains a critical bottleneck. Enhanced sorting systems can increase the fraction of waste recoverable for RDF production, minimising the residual waste sent to landfills.
- Infrastructure and logistics development: Efficient RDF valorisation requires robust infrastructure for waste collection, processing, and transportation. Establishing region-specific RDF hubs could ensure a steady supply of RDF to cement plants.
- Ensuring compliance with emission norms: The use of RDF in cement kilns must adhere to stringent emission standards to avoid secondary environmental impacts, such as the release of dioxins or heavy metals during combustion.
- Public and industrial awareness: Educating municipalities, industries, and communities about the environmental benefits of RDF can foster greater adoption and collaboration in waste management initiatives.

13.6.3 Environmental Concerns: Dioxin Emissions from RDF Co-processing in Cement Kilns

One of the primary environmental concerns related to the co-processing of RDF in cement kilns is the potential formation and emission of polychlorinated dibenzo-p-dioxins and dibenzofurans (PCDD/Fs), commonly referred to as dioxins. These are highly toxic, persistent organic pollutants (POPs) that can have serious environmental and health implications, even at trace concentrations.

Mechanism of dioxin formation: dioxins can be formed during combustion through two principal mechanisms (Pacific Air & Environment 2002):

- Formation from precursor—where chlorinated aromatic hydrocarbons (e.g., chlorobenzenes, chlorophenols), present in the RDF feedstock or generated during incomplete combustion, transform into PCDD/Fs in the post-combustion cooling phase.
- Formation "de novo" synthesis—where carbonaceous particles containing chlorine and catalytic metals (e.g., copper) facilitate the formation of dioxins on particulate surfaces in the temperature range of 250–450 °C, particularly in downstream equipment such as air pollution control devices or flue gas ducts.

These critical temperatures are not typically encountered in the main combustion zone of a cement kiln but may occur in the cooling sections or filters, making process control in these areas essential.

Cement kilns, especially rotary kilns, offer unique thermal and residence time conditions that are inherently suited for the complete destruction of organic pollutants (Pacific Air & Environment 2002):

- Flame temperatures exceed 2,000 °C.
- Residence times are longer compared to conventional incineration plants.
- Combustion occurs in a counter-current flow setup, ensuring thorough mixing of kiln feed and exhaust gases.
- Oxygen-rich conditions reduce the likelihood of incomplete combustion.

These conditions contribute to almost complete thermal destruction of organic matter, including dioxin precursors.

Data from 37 cement kilns in Germany (VDZ 2022) show that in 65% of the measurements, no PCDD/Fs were detected in the clean flue gas. Importantly, these findings demonstrate that:

- PCDD/F emissions are not significantly influenced by the type of input material (including RDF).
- Emissions can be effectively controlled by proper process design and flue gas management (Fig. 13.10).

This suggests that, with proper operational practices and flue gas cooling strategies, cement kilns can safely co-process RDF without contributing to elevated dioxin emissions.

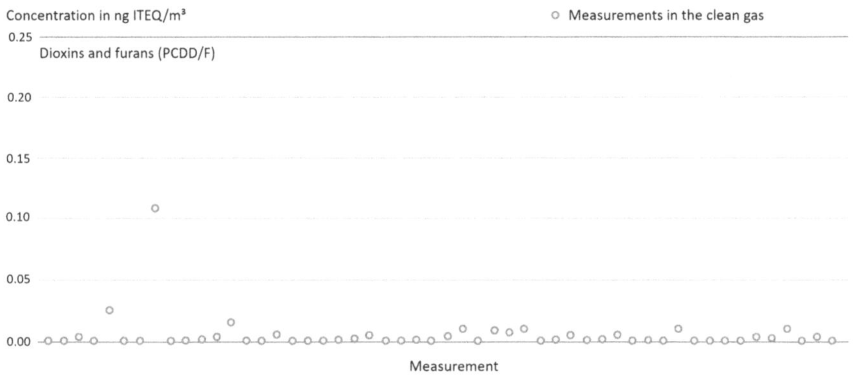

Fig. 13.10 Measured values in the clean exhaust gas. In 65 cases no PCDD/F was detected. (VDZ 2022)

13.6.3.1 Regulations in India and Europe, Dioxin Emissions Are Tightly Regulated

- The European Industrial Emissions Directive (IED) mandates a maximum emission limit of 0.1 ng I-TEQ/m^3 (International Toxic Equivalents) for PCDD/Fs (European Commission 2010).
- India follows guidelines issued by the CPCB, which also reference this limit for waste-to-energy and co-processing facilities, including cement kilns (CPHEEO 2018).

13.6.3.2 Compliance Requires

- Continuous monitoring of flue gas quality.
- Maintenance of optimal combustion parameters.
- Proper temperature management in pollution control devices to avoid the "de novo" formation window.

While the theoretical risk of dioxin formation exists due to the heterogeneous and chlorinated nature of RDF, cement kilns—when properly operated—provide a highly controlled and high-temperature environment that ensures near-complete destruction of PCDD/Fs. When complemented by post-combustion flue gas cooling and emission monitoring, the use of RDF in cement kilns poses minimal risk of dioxin release. Nonetheless, robust regulatory oversight and adherence to best available techniques (BAT) are crucial to ensuring public and environmental safety.

13.7 Conclusion

The valorisation of RDF through landfill mining offers a transformative approach to addressing India's growing waste management challenges while contributing to energy security and environmental sustainability. As highlighted in this paper, RDF production leverages the untapped potential of landfill mining, converting non-recyclable waste fractions into a valuable alternative fuel source for energy-intensive industries like the cement industry.

With an estimated 1.72 million tonnes of RDF extractable from India's aged MSW annually, if all cement plants replace up to 30% of heat consumption in cement kilns, it will require 20.76 million tonnes of RDF (at 3000 kcal/kg of RDF or other alternative fuels). It means there is not enough waste to fulfil the demand. Bridging this gap requires, tapping into additional biomass resource in India.

RDF valorisation also delivers significant environmental benefits, such as reducing greenhouse gas emissions, mitigating methane release, and minimising groundwater contamination. For instance, replacing coal with RDF at a 30% TSR in cement plants could reduce annual CO_2 emissions by approximately 9.38 million tonnes, representing an 11.3% decrease compared to baseline coal usage from Indian cement industry. Moreover, this approach supports the Swachh Bharat Mission by diverting waste from open burning, reducing landfill dependency, and promoting circular economy principles.

However, realising this potential requires addressing critical challenges, including improving waste segregation efficiency, developing robust RDF processing infrastructure, ensuring compliance with emission norms, and fostering public and industrial awareness. Strategic investments in these areas, combined with supportive policies and technological advancements, can unlock the full potential of RDF as a sustainable fuel source in India.

In conclusion, the transition to RDF valorisation from landfill mining represents a win-win scenario for waste management and energy production. It not only mitigates the environmental impacts of landfill waste but also supports industrial decarbonisation, contributing to a cleaner, greener, and more sustainable future for India. Continued research, stakeholder collaboration, and policy reforms will be essential in driving this transformation forward.

References

Bhandare VM, Bhosale SM, Patil YM (2024) Legacy waste management: a case study of Pune City. Int Res J Eng Technol (IRJET) 11(1):565–572

Central Pollution Control Board (CPCB) (2019) Guidelines for disposal of legacy waste (old municipal solid waste). Available from: https://cpcb.nic.in/uploads/LegacyWasteBiomining_guidelines_29.04.2019.pdf

Central Pollution Control Board (CPCB) (2022) Annual report on the status of implementation of Municipal Solid Waste (MSW) rules. Retrieved January 8, 2025, from https://cpcb.nic.in/status-of-implementation-of-solid-waste-rules/

Central Public Health and Environmental Engineering Organisation (CPHEEO) (2018) Guidelines on usage of refuse-derived fuel (RDF) in various industries

European Commission (2010) Directive 2010/75/EU of the European Parliament and of the Council of 24 November 2010 on industrial emissions (integrated pollution prevention and control). Off J Eur Union L334:17–119. https://eur-lex.europa.eu/legal-content/EN/TXT/?uri=CELEX%3A32010L0075

European Investment Bank (EIB) (2024) Managing refuse-derived and solid recovered fuels: best practice options for EU countries. EIB Sector Paper, 2024/01

Ghosh A, Kartha SA (2023) Composition and characteristics of excavated materials from a legacy waste dumpsite: potential of landfill biomining. Environ Res Technol 6(2):108–117. https://doi.org/10.35208/ert.1245574

India Brand Equity Foundation (IBEF) (2025) Indian cement industry analysis. India Brand Equity Foundation. Retrieved January 8, 2025, from https://www.ibef.org/industry/cement-presentation

Intergovernmental Panel on Climate Change (IPCC) (2006) 2006 IPCC guidelines for national greenhouse gas inventories

International Organization for Standardization (ISO) (2021) ISO 21640:2021: solid recovered fuels—specifications and classes. Retrieved January 8, 2025, from https://www.iso.org/standard/71309.html

Lechtenberg D (2007) Rebuilding Beirut with alternative fuels. Global Fuels Magazine

Lechtenberg D, Diller H (2012) Alternative fuels and raw materials handbook, vol 1. MVW Lechtenberg & Partner. ISBN: 978-3-7640-0551-1

Pacific Air & Environment (2002) Sources of dioxins and furans in Australia: air emissions (Revised ed.). Environment Australia. https://www.dcceew.gov.au/sites/default/files/documents/dioxinsources.pdf

Saha PK, Karstensen KH (2017) Coprocessing of alternative fuels & resources in Indian cement industry: baseline and potential. In: Global waste management symposium 2017

Somani M, Datta M, Ramana GV, Sreekrishna TR (2018) Investigations on fine fraction of aged municipal solid waste recovered through landfill mining: case study of three dumpsites from India. Waste Manag Res. https://doi.org/10.1177/0734242X18782393

Statista. India (2025) Estimated total population from 2020 to 2030, Statista, https://www.statista.com/statistics/263766/total-population-ofindia/

VDZ (2022) Umweltdaten der deutschen Zementindustrie 2021. Duesseldorf. Available at: www.vdz-online.de/wissensportal/publikationen/umweltdaten-der-deutschen-zementindustrie-2021

Vollmann F (2019) After the tragic landslide that killed 116, Koshe landfill in Addis Ababa is safer. United Nations Human Settlements Programme (UN-Habitat) website. Retrieved January 22, 2025, from https://unhabitat.org/news/05-jul-2019/after-the-tragic-landslide-that-killed-116-koshe-landfill-in-addis-ababa-is-safer

Zement Kalk Gips (ZKG) International (2015) The Indian cement sector: technological status and prospects. Retrieved from https://www.scribd.com/document/19424358/Rebuilding-Beirut-With-Alternative-Fuels

Part IV
Implementation Aspects

Chapter 14
Unlocking the Potential of Landfill Mining in India: A Circular Approach to Waste Management

Anusree Nalladiyil, B. Prathima, and G. L. Sivakumar Babu

14.1 Introduction

The rapid urbanization and population growth in India have led to significant increase in municipal solid waste (MSW) generation, leading to the increase in dumpsites or landfills, posing significant environmental and public health challenges (Ghosh and Kartha 2024; Kaza et al. 2018). Traditionally, landfills have been considered a primary waste disposal method. According to the Central Pollution Control Board (CPCB), India generates approximately 170,339 tonnes per day (TPD) of waste, of which 156,449 TPD is collected, achieving a 92% collection efficiency. However, only 54% of the total waste generated (91,511 TPD) undergoes scientific processing, while 24% (41,455 TPD) is disposed of in landfills (CPCB 2016). The remaining 22% (37,373 TPD) represents a critical gap in daily MSW management, indicating inefficiencies in waste treatment infrastructure and processing capacity. Resulting in unscientific dumping of MSW, which poses a significant threat to human health and the surrounding ecosystem due to leachate

A. Nalladiyil (✉)
Centre for Sustainable technologies, Indian Institute of Science Bangalore, Bengaluru, Karnataka, India
e-mail: anusreen@iisc.ac.in

B. Prathima
Centre for Sustainable technologies, Indian Institute of Science Bangalore, Bengaluru, India

Department of Civil Engineering, B M S College of Engineering, Bengaluru, India

G. L. S. Babu
Department of Civil Engineering, Indian Institute of Science Bangalore, Bengaluru, Karnataka, India

M. Somani et al. (eds.), *Landfill Mining of Legacy MSW: Global Insights and Indian Experience*, Springer Transactions in Civil and Environmental Engineering, https://doi.org/10.1007/978-981-95-3959-8_14

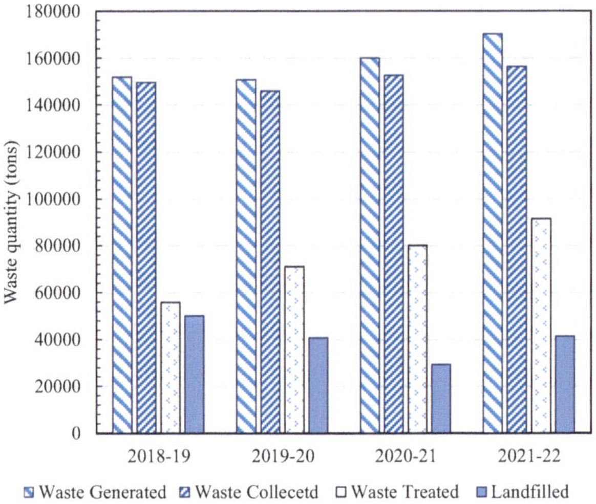

Fig. 14.1 Solid waste management status in India from 2018 to 2022 (CPCB annual reports, 2018 to 2022)

generation, gas emissions, fire hazards, and surface and subsurface contamination (Goli and Singh 2023; Singh et al. 2022; Singh and Budarayavalasa 2024).

While there has been notable progress in waste treatment technologies, the rise in waste generation has outpaced improvements in waste collection and processing. Between 2018–19 and 2021–22, waste treatment capacity improved by 16.95%, driven by advancements in waste-to-energy initiatives, composting, and material recovery (Fig. 14.1). However, during the same period, waste collection efficiency declined by 6.7%, suggesting that infrastructure expansion has not kept up with increasing waste generation. Additionally, despite an overall 8.64% reduction in landfill dependency, a 5.95% increase in landfill disposal was recorded in 2021–22, signalling an urgent need for enhanced waste diversion policies and decentralized waste management solutions. Despite this shift, a substantial portion of India's legacy waste still present in the old dumpsites and poorly managed these sites contributing to severe environmental hazards, including groundwater contamination, greenhouse gas emissions, and land degradation (CPCB 2019). Therefore, there is critical need for effective legacy waste management in India.

To address these challenges, landfill mining—often termed "biomining"—has gained traction, particularly in India (Deendayal and Kasinathan 2022; Kaartinen et al. 2013; Kurian et al. 2007; Wagner and Raymond 2015), approach aims to rehabilitate valuable land, safeguard groundwater, and recover secondary resources, including combustible, compostable, and recyclable materials, from landfills. However, challenges such as waste characterization, cost-effectiveness, and treatment of excavated waste need to be addressed. Despite these challenges, landfill mining offers technological opportunities to extract valuable resources and promote sustainable waste management. This situation underscores the need for a systematic approach to landfill mining and dumpsite remediation as a means to reclaim valuable land, recover resources, and mitigate the long-term environmental impacts of accumulated legacy waste (Gusca et al. 2015). Additionally, integrating circular economy principles into landfill mining strategies can facilitate the recovery of recyclables, energy, and secondary raw materials, transforming waste management into a more sustainable and economically viable sector (Singh and Budarayavalasa 2024). This highlights the urgent need for policy interventions to enhance landfill mining activities more sustainable and circular. Therefore, this chapter aims to identify the existing policy landscape and its role in facilitating landfill reclamation. The study identifies gaps in governance and policy implementation to enhance the efficiency of landfill mining operations Furthermore, it explores the integration of circular economy models into landfill mining practices to maximize resource recovery, minimize environmental and health risks, and advance sustainable waste management solutions.

14.2 Initiatives by Indian Government

India's approach to waste management and landfill mining aligns with its broader circular economy goals, emphasizing waste-to-resource conversion, landfill remediation, and sustainable urban planning. India has launched several initiatives to promote sustainable waste management, landfill remediation, and resource recovery as part of its broader environmental and urban development policies. Key programmes include:

- The Swachh Bharat Mission (SBM), initiated in 2014, intends to attain comprehensive scientific waste management in both urban and rural regions. The mission emphasizes solid waste management, source segregation, and waste-to-energy projects. In its second phase (SBM 2.0), initiated in 2021, landfill restoration and legacy waste management have been emphasized to reclaim land and mitigate environmental risks.
- The Clean India Mission is consistent with SBM and promotes public involvement in cleanliness initiatives, sustainable sanitation methods, and solid waste management. It has been essential in enhancing awareness regarding waste reduction, segregation, and recycling.
- The AMRUT programme, initiated in 2015, aims to enhance urban infrastructure, specifically water supply, sewage systems, and green areas in 500 communities. It incorporates waste management by enhancing wastewater treatment and advocating for localized waste processing alternatives.
- India has implemented various policies under its National Action Plan for waste Management, such as Extended Producer Responsibility (EPR) for plastics and electronic waste, the promotion of circular economy ideas, and the encouragement of waste-to-energy initiatives.

- The Swachh Bharat Mission-Gramin (SBM-G) effort facilitates the transformation of biodegradable waste, particularly from livestock and organic sources, into biogas, compost, and biofertilizers to foster a circular economy through GOBAR-Dhan Scheme (Galvanizing Organic Bio-Agro Resources Dhan).
- Landfill Mining and Remediation State Initiatives: Several cities, including Indore, Delhi, and Pune, have implemented landfill mining projects to recover valuable materials, reduce landfill footprints, and reclaim land for urban development. The success of Indore's landfill remediation under SBM 2.0 has set a precedent for other cities. Various Indian states have launched their own initiatives, such as Maharashtra's Swachh Maharashtra Mission, Tamil Nadu's Solid Waste Management Strategy, and Kerala's Haritha Keralam Mission, focusing on waste reduction, composting, and decentralized waste management.

These programmes and policy directives have increasingly emphasized the sustainable management of waste materials, recognizing landfill mining as an opportunity to advance sustainability goals. The implementation of these programmes is critical for mitigating environmental hazards, improving public health, and achieving sustainable urban development. The introduction of the Solid Waste Management (SWM) Rules, 2016, which superseded the 2000 edition and offers a stronger framework for waste management in India, marked a major turning point in this endeavour. The legal mandate under Solid Waste Management Rules, 2016 rule 15 on existing dumpsites states that, the local authorities and urban local bodies shall, "*Investigate and analyse all old open dumpsites for their potential of biomining and bioremediation and take necessary actions to bio-mine or bio-remediate the site.*" To support this clause and to facilitate effective reclamation of dumpsites, CPCB released the Guidelines for Disposal of Legacy Waste (Old Municipal Solid Waste) on direction of National Green Tribunal (NGT). Furthermore, the regulations emphasize the remediation of legacy landfills through bioremediation or engineered capping. Despite these advancements, challenges persist due to infrastructure gaps, financial constraints, and the continuous influx of waste into existing dumpsites, underscoring the need for robust implementation strategies, stakeholder participation, and sustainable waste management solutions.

In India, scientific management of dumpsites has been a major thrust area in urban India under the national flagship mission Swachh Bharat Mission Urban 2.0 (SBM-U 2.0). SBM-U 2.0 builds on the foundation of SBM-U 1.0 by targeting 100% waste processing through bioremediation, biomining, and decentralized treatment solutions. It enhances resource recovery by promoting co-processing in cement kilns and integrating the informal waste sector for improved material reuse. The mission leverages advanced technologies such as GIS-based waste tracking, AI-driven waste processing, and smart waste management systems to optimize operations. SBM-U 2.0, launched in 2021, aims to enhance urban sanitation and solid waste management, with a key focus on remediating approximately 2400 legacy landfill sites by 2025–2026 through landfill mining. This process involves excavating and processing waste to recover resources and reclaim land. However, significant challenges remain in achieving these targets. In cities with populations exceeding one million, 69 major landfill sites collectively occupy 3354 acres and contain approximately 1258 lakh metric tonnes of waste. While 35 sites have undergone partial remediation, only 1171 acres have been cleared, processing 475 lakh metric tonnes of waste. Figure 14.3 depicts the condition of dumpsites in Indian states and Union Territories, emphasizing the number of identified dumpsites, remediation progress, completed remediation efforts, and sanctioned remediation action plans.

Key highlights include the elimination of 16 crore metric tonnes of legacy waste, a push for scientific waste management through waste-to-energy, composting, and material recovery, and a stronger focus on circular economy principles, including plastic waste reduction. It also emphasizes decentralized waste processing through small-scale biogas plants and composting units, along with increased financial support and incentives for zero-landfill models. With digital interventions and stricter waste diversion policies, SBM-U 2.0 shifts from basic sanitation and waste collection to sustainable waste processing and landfill reclamation, making it a comprehensive and future-ready initiative. These initiatives represent a crucial advancement in revolutionizing India's waste management framework by tackling current waste issues and facilitating resource recovery and environmental rehabilitation. Consequently, evaluating the present condition of dumpsites and landfill mining operations is essential for comprehending the advancement and future direction of these endeavours.

14.3 Current Status of Dumpsites and Landfill Mining Activities

Landfill reclamation has become an urgent necessity due to the legal and environmental imperatives by Indian government and the NGT Act has played a crucial role in shaping India's landfill policies by enforcing strict waste disposal regulations and mandating scientific landfill management practices. The NGT has issued several orders directing municipalities to phase out unscientific and unregulated dumpsites while ensuring compliance with the SWM Rules, 2016. The NGT has emphasized the need for biomining and bioremediation of old landfills, directing state governments to reclaim land occupied by legacy waste through scientific methods. This aligns with India's push towards landfill mining and resource recovery.

Figure 14.2 indicates the status of dumpsites in India over the four-year period 2018–2022, significant progress was made in dumpsite remediation, with the total number of dumpsites declining from 3135 in 2018–2019 to 2452 in 2021–2022, indicating large-scale reclamation efforts. The overall count of dumpsites has stayed rather constant, suggesting that continuous waste management initiatives are under action even if additional sites could be developed. Especially, the quantity of dumpsites reclaimed or capped has exhibited a consistent rise, which reflects developments in waste management projects. The number of reclaimed or capped dumpsites

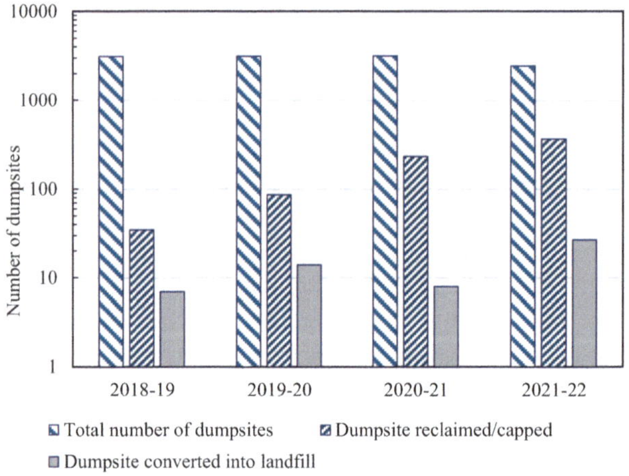

Fig. 14.2 Status of dumpsites in India from 2018 to 2022 (CPCB annual reports, 2018–2022)

increased tenfold, from 35 in 2018–19 to 366 in 2021–22, reflecting a strong push towards dumpsite closure and environmental restoration. However, dumpsite conversion into engineered landfills showed fluctuating progress, initially increasing from 7 in 2018–19 to 14 in 2019–20, dropping to 8 in 2020–21, and then surging to 27 in 2021–22. This suggests an evolving approach in waste disposal, with a growing emphasis on sustainable landfill management. The decline in the total number of dumpsites in 2021—22, despite a steady increase in the previous years, highlights the impact of accelerated remediation initiatives. However, ensuring long-term sustainability requires continued investments in waste diversion, landfill rehabilitation, and circular economy-driven waste management strategies to prevent the formation of new dumpsites.

Figure 14.3 shows the progress of dumpsite remediation across Indian states. Maharashtra (236), Tamil Nadu (296), Uttar Pradesh (223), Karnataka (213), and Madhya Pradesh (167) have the highest number of dumpsites, reflecting their large populations and high waste generation rates which also have major metro cities. Gujarat, Maharashtra, and Tamil Nadu report the highest number of remediated dumpsites, with 118, 114, and 117 sites completed, respectively. However, states like Uttar Pradesh and Karnataka but show relatively slow progress in remediation. Several states, including Delhi, Jammu & Kashmir, and Punjab, have a considerable number of remediation projects currently in progress. Meanwhile, states such as Nagaland and Meghalaya report no remediation activity. West Bengal (117 dumpsites, 0 completed, 7 in progress, no approved action plan) stands out as a region with very little remediation effort. Jammu & Kashmir has a high number of approved plans (25) but no completed remediation, suggesting early-stage efforts. The data also highlight discrepancies between the number of approved remediation action plans and actual implementation, indicating possible delays in execution. Overall, while several states have made significant strides in landfill reclamation, a

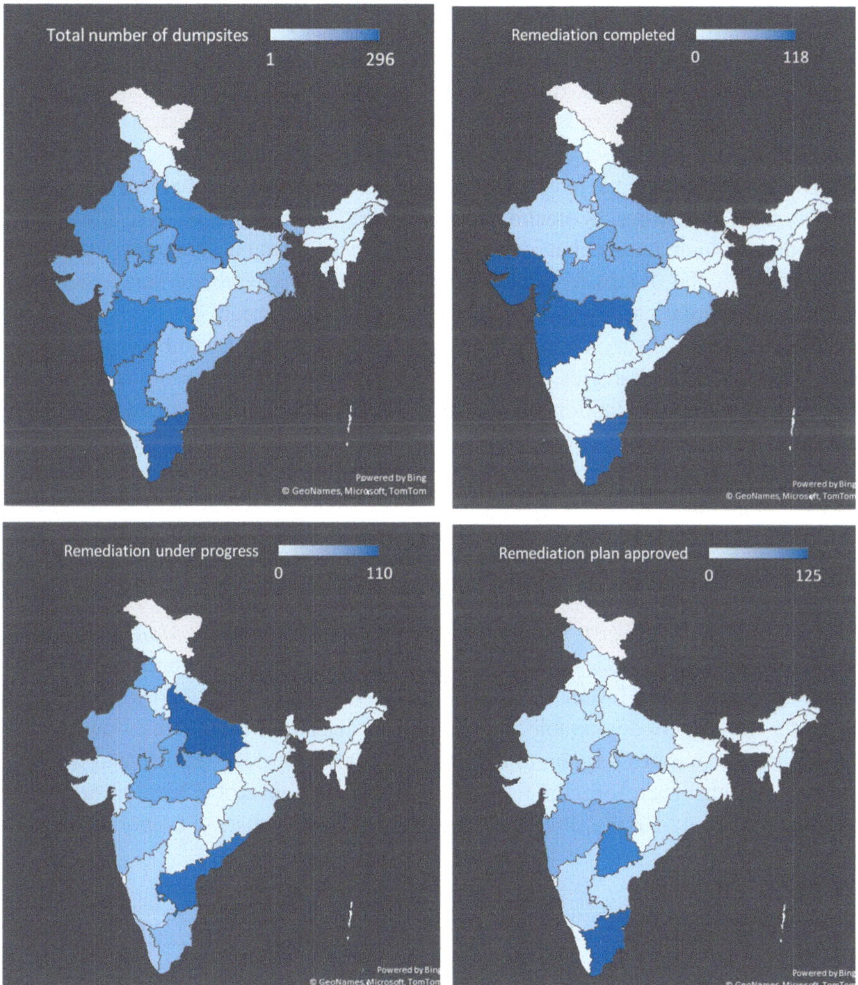

Fig. 14.3 Number of dumpsites and remediation status across different state. (https://sbmurban.org/swachh-bharat-mission-progess, Accessed second March 2025)

comprehensive and accelerated approach is necessary to address the remaining untreated sites effectively.

States like Maharashtra, Karnataka, Uttar Pradesh, and Tamil Nadu possess a greater number of dumpsites, whilst smaller states and Union Territories such as Goa, Puducherry, and Ladakh have fewer. Certain states, such as Maharashtra and Gujarat, have achieved notable advancements in remediation efforts, while others, despite possessing multiple sanctioned remediation plans, exhibit sluggish execution, as evidenced in West Bengal and Madhya Pradesh. Karnataka, Tamil Nadu, and Rajasthan demonstrate a measured strategy with continuous remediation efforts.

14.4 Landfill Mining Operation

Landfill mining operations require a structured process. In old and uncontrolled dumpsites, where no records of disposal activities exist, geophysical surveys are essential for assessing waste deposits (Mandpe et al. 2019). These surveys help determine the depth and extent of waste layers and identify saturated leachate zones. Based on these findings, geotechnical surveys are conducted to analyse site stratigraphy and evaluate the mechanical stability of waste during excavation. A thorough assessment of the dumpsite's environmental conditions is also crucial, including leachate accumulation, biogas production potential, and current emission rates. This information is necessary for selecting appropriate excavation technologies, planning pre-treatment strategies, and implementing site-specific measures to prevent environmental contamination and ensure worker safety. To maximize resource recovery from excavated waste, analysing the grain-size distribution and fractional composition of bulk waste samples is essential. In the absence of historical data on waste composition, systematic sampling must be conducted at multiple locations within the dumpsite at regular intervals to obtain a representative characterization (Krook et al. 2012).

The standard protocol outlined in the *Guidelines for Disposal of Legacy Waste (Old Municipal Solid Waste)* and the *Legacy Waste Management Manual* is shown in Fig. 14.4. The legacy waste is excavated using frontend or backend loaders, breaking down waste layers for further processing and transported to designated areas for drying. Biodegradable waste, such as discarded food, fruit, flowers, and garden waste, requires oxygen to decompose in an odourless manner without producing leachate. Therefore, the first step in stabilizing anaerobic legacy waste is to expose as much of it as possible to air. This step allows leachate to evaporate through

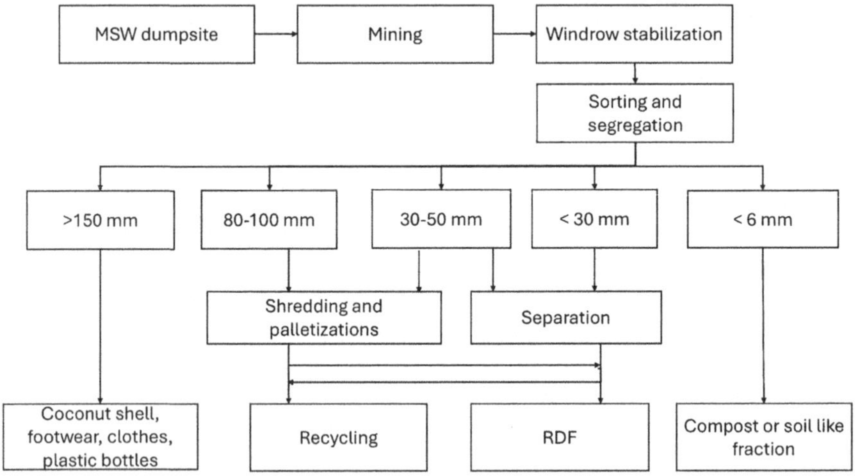

Fig. 14.4 Flow chart of landfill mining operations (CPCB Guidelines)

solar exposure and ensures the release of entrapped methane from the waste heap. Once dried, trommel screens are used to separate the waste into different fractions. The screened materials are classified into refuse-derived fuel (RDF), non-recyclable rejects, fine soil-like fractions (<4 mm), and metals or other recyclables. However, landfill mining operations face significant challenges during the rainy season due to increased moisture content in the excavated waste, making drying and screening more difficult.

14.5 Typical Equipment and Vehicle Used in Landfill Mining Operations

Landfill mining operations require a range of specialized vehicles and equipment to efficiently excavate, screen, and process waste materials. The selection of machinery depends on the scale of the operation, waste composition, and the intended use of the recovered materials. Landfill mining operation, vehicles and equipment utilized to facilitate the excavation and processing of legacy waste is shown in Fig. 14.5. Backhoe loaders were used for excavation, lifting, and loading waste into screening units or transport vehicles, offering versatility and efficiency. Mobile trommel screens played a crucial role in on-site waste separation, classifying different fractions of waste based on particle size to enhance material recovery. Additionally, fixed trommel screens were also installed at centralized processing facilities for large-scale waste segregation, enabling continuous screening and sorting of materials for further treatment or recycling. Usually, heavy-duty trucks were employed to transport excavated waste from the landfill to processing or disposal sites, ensuring smooth material flow. Compactors were used to reduce waste volume before

Fig. 14.5 Vehicle used in the landfill mining operation (**a**) Truck (**b**) Compactors (**c**) Excavator and Mobile Trommel (**d**) Backhoe loaders (**e**) Fixed trommels

transportation, optimizing space utilization and aiding in the stabilization of residual waste. These vehicles and equipment were integral to the landfill mining operation, enhancing efficiency, and enabling resource recovery as part of sustainable waste management efforts. The use of specialized vehicles and equipment in landfill mining operations, involves significant capital and operational costs. However, their efficiency in excavation, waste processing, and material recovery can offset these expenses through optimized resource utilization. Mobile trommel screens offer an economical advantage by enabling on-site waste segregation, reducing dependency on centralized processing facilities and minimizing transportation expenses. While fixed trommels require higher initial investment and infrastructure, they enhance operational efficiency by allowing continuous waste screening, improving the quality of recovered materials such as soil, recyclables, and RDF. Effective planning and optimization of equipment use can further enhance cost efficiency, making landfill mining a financially and environmentally sustainable waste management strategy.

14.6 Circular Economy Model for Resource Recovery and Disposal

The circular economy framework for landfill mining emphasizes resource recovery, material repurposing, and waste minimization to reduce the need for re-landfilling. By effectively sorting and processing waste fractions, valuable materials can be reintegrated into industrial and construction applications, reducing environmental impact and promoting sustainability. In Indian landfill mining studies, landfilled waste was composed primarily of soil-like material (61.91%), which can be reused for landfill cover, land reclamation, and construction fillers after contamination assessment. Plastics (8.49%) can be processed into refuse-derived fuel (RDF) or recycled into new plastic products. The stone fraction (22.30%), along with glass (1.76%) and metals (1.87%), can be repurposed for construction applications, reducing the demand for virgin aggregates. Wood (2.30%) can be repurposed for bioenergy or composite materials, while textiles (2.88%) can be used for RDF production (Fig. 14.6).

The potential for resource recovery or disposal from dumpsite material largely depends on several key factors. The composition of waste within the dumpsite plays a crucial role, as different materials have varying recovery potentials. Additionally, the quality of the waste, including its level of degradation and contamination, affects its suitability for reuse or recycling. The effectiveness of physical characterization and the availability of efficient segregation and treatment technologies further determine how well different material fractions can be processed. Lastly, the feasibility of resource recovery is influenced by the availability of suitable end-use applications or markets for the recovered materials. By minimizing the volume of unrecoverable waste, only a small fraction of residual material will need to be re-landfilled, extending landfill lifespan and reducing environmental pollution. Remediated

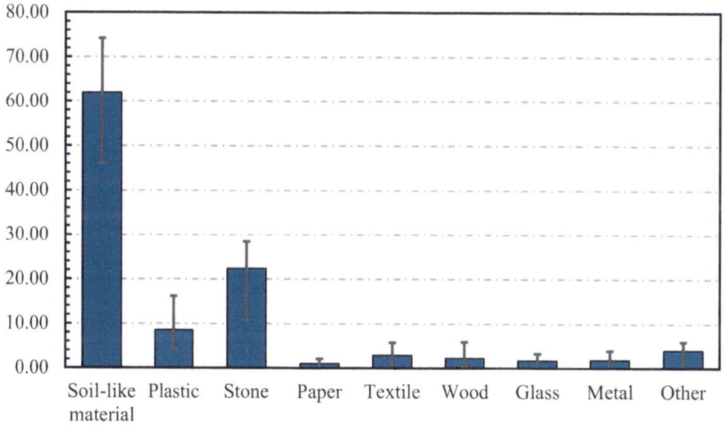

Fig. 14.6 Landfill mined waste composition for Indian landfill mining studies ($n = 10$)

landfill sites can then be repurposed for urban development, green spaces, or renewable energy projects like solar farms. To enhance landfill mining viability, Extended Producer Responsibility (EPR) policies should encourage industries to use recycled materials. Establishing market linkages for recovered materials, along with quality certification systems, will drive demand. Financial incentives such as green bonds and carbon credits can further support large-scale landfill remediation projects.

Depending on the waste composition and its properties, recovered fractions can be used for waste-to-energy applications, soil enhancement, embankment filling, or landfill cover. Plastic, textiles, etc. can be processed into refuse-derived fuel (RDF) for energy recovery in industries. Refuse-derived fuel (RDF) is one of the key outputs of landfill mining, with potential applications in energy recovery. However, the RDF potential from excavated dumpsite waste often falls below the minimum requirement of 1500 kcal/kg specified in Indian standards for use in cement kilns. To enhance its usability, preprocessing strategies such as high-calorific-value material mixing and quality improvement measures need to be implemented. Additionally, logistical aspects such as transportation to cement kilns, brick factories, or waste-to-energy plants must be considered to ensure cost-effective utilization.

The soil-like fraction recovered from landfill mining, often referred to as biosoil, constitutes a major portion of the excavated material (Datta et al. 2021; Somani et al. 2018, 2022). This fraction can be repurposed for various applications, including embankment filling, landfill cover, and even soil enhancement in specific cases. The suitability of biosoil for reuse depends on its physical and chemical characteristics, such as texture, organic matter content, and contaminant levels. Proper treatment and stabilization techniques may be required to ensure compliance with environmental regulations before its application. Further research is ongoing to explore the potential of utilizing biosoil in agriculture, reclamation projects, and other sustainable land applications.

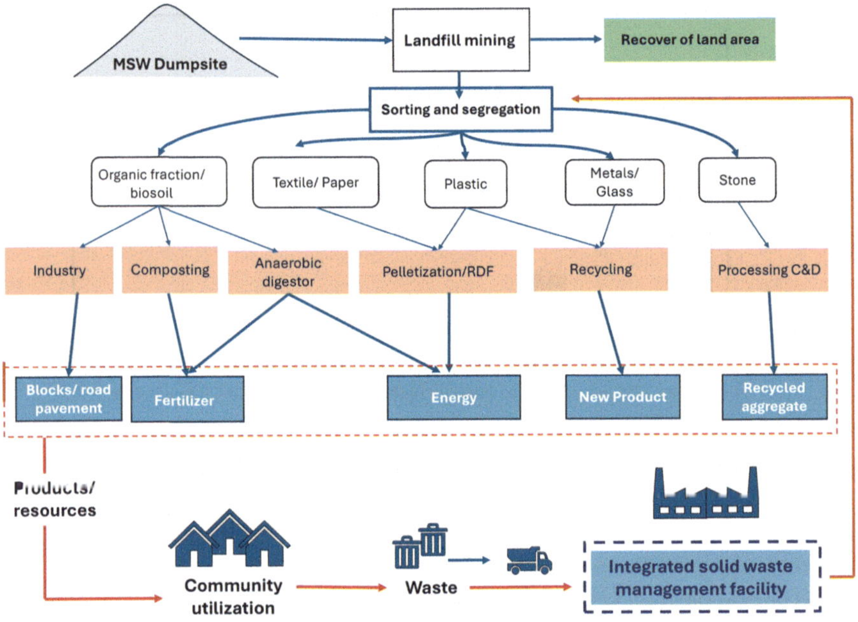

Fig. 14.7 Circular economy framework for the management of legacy waste

Inert fractions, including aggregates, stones and soil, can be repurposed for construction activities, meeting quality standards for use in building materials. Fine fractions from landfill mining, especially from industrial waste, offer potential for recovering valuable metals and aggregates. The coarser materials from construction and demolition (C&D) waste can be converted into recycled aggregates, which are useful for road subbase layers, landfill capping, and drainage systems. The fine soil-like fraction can serve as filler material or landfill cover, further supporting sustainable waste management. The circular economy approach to legacy waste management enables resource recovery from dumpsites, reducing environmental impact and promoting sustainability (Fig. 14.7). The circular economy framework for landfill mining focuses on maximizing resource recovery while minimizing the volume of waste that requires re-landfilling. This approach ensures that valuable materials extracted from dumpsites are repurposed, reducing environmental burdens and promoting sustainable waste management.

14.7 Value Addition of Landfill Mining

Landfill mining was initially driven by the need for land reclamation and remediation. In recent years, however, the focus has shifted towards recovering valuable materials and resources. Studies assessing landfill mining projects have highlighted both environmental and economic advantages, particularly in terms of land restoration (Deendayal and Kasinathan 2022; Jain et al. 2023; Wagner and Raymond 2015). Jain et al. (2014) conducted a life cycle impact assessment (LCIA) of landfill mining and found that emissions from diesel production and use in mining equipment were relatively minor compared to the emissions from a "do nothing" scenario, where the landfill remained undisturbed. Their study demonstrated that relocating waste led to a significant reduction in global warming potential (GWP), even when mining was assumed to occur after 30 years of waste decomposition, a stage when gas production generally declines. Cheela et al. (2021) during the LCIA of legacy waste found that the application of landfill-recovered soil-like material as compost provided significantly greater environmental benefits compared to a no-mining scenario. The implementation of mining processes and soil application resulted in an 80% reduction in GWP. Additionally, the avoidance of virgin metal extraction in manufacturing contributed to further environmental savings within the GWP category.

Among the recovered materials, soil-like fractions account for the largest share, typically comprising 50–60% of the excavated waste. While this material is primarily repurposed as landfill cover, its potential for broader land applications remains under investigation. Managing landfills with vast waste accumulations poses significant challenges for municipalities, where economic feasibility remains the primary consideration in designing mining strategies. Environmental benefits, though acknowledged, are often assessed qualitatively. As sustainability considerations gain traction, municipalities are striving to implement landfill mining systems that balance environmental responsibility, economic practicality, and social acceptance. Currently the market value of excavated waste is very limited. Proper disposal of the recovered material is a challenge which is leading to the accumulation of the waste in the landfill mining site. Additionally, the market for recovered RDF is also low, due to low calorific value. A preprocessing of the RDF to increase the calorific value can be carried out. Analysing volatile solids, ash content, and energy value of legacy waste, give the necessity of implementing pre-cleaning techniques—cleaning, drying, and sorting—to eliminate surface impurities from the recovered waste (Cheela et al. 2021). Additionally, advanced treatment technologies such as gasification, pyrolysis, and hydrothermal carbonization offer promising pathways for utilizing co-mingled feedstocks. Thereby landfills can act as resource reservoirs, presenting significant potential for future energy recovery and sustainable waste management. Additionally, the rehabilitation of old dumpsites offers environmental benefits, including reduced groundwater and air contamination and methane explosion risks. In some instances, reclaimed land has been repurposed for public parks or other developments, demonstrating the potential for sustainable landfill mining projects.

The economic feasibility of landfill mining operations also depends on the cost of vehicles and equipment but also on labour expenses. The use of heavy machinery significantly improves operational efficiency and reduces dependence on manual labour. However, landfill mining remains labour-intensive, particularly in tasks such as waste segregation, material handling, and quality control of recovered resources. The overall expenditure includes the cost of both skilled and unskilled labour, including machine operators and supervisors. A key challenge is the availability of skilled labour capable of handling specialized equipment and overseeing operations effectively. In India, manual segregation was employed for recovering metal scrap and other recyclables, highlighting the continued reliance on human labour for precision sorting. In India, where labour costs are lower compared to developed nations, manual segregation often complements mechanical sorting, enhancing recovery rates of valuable materials. While mechanization reduces processing time and improves overall efficiency, an optimal balance between labour and equipment is essential to ensure cost-effective operations.

14.8 Case Studies

Landfill mining has arisen as a strategic method for tackling the issues associated with the accumulation of legacy garbage in metropolitan areas. Landfill mining corresponds with circular economy ideas and sustainable urban development by harvesting valuable materials, reclaiming land, and mitigating environmental hazards. Numerous cities in India have initiated landfill rehabilitation initiatives, showcasing a variety of approaches, legislative interventions, and technological innovations in garbage recovery and disposal.

This section summarizes case studies of landfill mining efforts in several Indian cities, emphasizing their implementation approaches, obstacles, environmental and economic ramifications, and exemplary practices (Table 14.1). Through the examination of these real-world instances, stakeholders can acquire insights into efficacious cleanup procedures, resource recovery potential, and policy frameworks that may be reproduced or scaled for forthcoming landfill mining initiatives

14.9 Challenges in Landfill Mining

Landfill mining presents several challenges, primarily due to the heterogeneous nature and contamination levels of excavated waste. The recovery of valuable materials is often hindered by high contamination, heavy metal presence, and the need for extensive processing. RDF often contain high ash and chlorine levels, reducing their energy potential and increasing operational costs (Chandana et al. 2020; Singh and Chandel 2020). Similarly, recyclable materials require intensive cleaning or washing, demanding additional investment in effluent treatment plants. Soil-like fractions typically lack sufficient concentration of recoverable metals, limiting their economic value. Compost derived from landfill mining often contains heavy metals exceeding permissible limits, making its use restricted to non-edible crops after expensive soil remediation (Chandana et al. 2020; Hölzle et al. 2022; Somani et al. 2018). Soil recovery is further complicated by high concentrations of heavy metals and soluble salts, requiring costly and time-intensive treatment. The use of

Table 14.1 Case studies of completed and ongoing landfill mining sites

Category	Bhopal	Chandigarh	Indore	Nashik	Tiruchirappalli
Location	Bhanpur dumpsite, Bhopal	Daddumajra dumpsite, Chandigarh	Devguradia dumpsite, Indore	Khatprakalp dumpsite, Nashik	Ariyamanglam dumpsite, Tiruchirappalli
Project duration	Jan 2018–June 2021 + 5 years monitoring	Oct 2019–(ongoing)	2016–2019	June 2017–Jan 2018	April 2019–Ongoing
Total area covered	37 acres	20 acres	100 acres	6 acres	39 acres
Reclaimed land	21.07 acres (57%)	8 acres (ongoing)	Not available	6 acres	26 acres (phase I complete, phase II ongoing)
Legacy waste processed	11,10,000 cum	3,75,000 MT	15,00,000 MT	1,90,000 MT	760,000 cum (500,000 MT)
Total cost	₹42 crore (remediation) + ₹10 crore (O&M)	₹33 crore	₹16 crore	₹3.4 crore	₹52 crore (phase I)
Cost per Tonne	₹378 per cum	₹660 per MT	₹660 per MT	₹180 per MT	₹684 per MT
Execution model	PPP (public-private partnership)	On weight basis	Not available	On weight basis	Based on volume of waste processed
Waste processing output	RDF: 19,000 MT Soil enricher: 63,220 MT Stones/C&D: 1606 MT	68% biosoil 14% C&D waste 16% RDF	50,000 MT processed in phase I 1.5 lakh MT processed in phase II	21,890 cum soil used for capping Recyclables given to ragpickers	83,010 MT RDF 2,64,032 MT soil enricher
End use of recovered land	Planned for interstate bus terminal	Future use for scientific waste treatment	Planned for urban forestry and land development	Used for C&D waste processing and composting	Planned for urban forest development
Monitoring & compliance	5 years post-closure monitoring Third-party audit	Continuous groundwater testing CCTV surveillance	Daily monitoring by IMC CCTV surveillance (30 cameras)	CCTV surveillance GPS-tracked vehicles	Third-party monitoring Online SCADA-based monitoring

excavated materials as landfills cover demands thermal, biological, or chemical treatment to address issues related to high total dissolved solids and organic content (Esakku et al. 2003; Somani et al. 2018). Additionally, repurposing recovered materials for construction necessitates comprehensive geotechnical testing, further escalating costs. These challenges significantly impact the feasibility and economic viability of landfill mining projects.

Another major challenge is the limited availability of advanced infrastructure and technology, which is essential for efficient landfill mining operations. Moreover, the high upfront costs associated with specialized equipment, monitoring systems, and waste handling infrastructure add financial constraints. The requirement for skilled personnel and training programmes further emphasizes the need for capacity building to ensure successful implementation. Operational efficiency is also impacted by waste composition variability and high moisture content, which reduce the consistency and effectiveness of the processes. Furthermore, the delayed economic returns and weak market demand for mined materials create financial uncertainty, leading to the piling up of excavated waste at the site and hindering the process. Addressing these barriers is crucial to enhancing the feasibility of biomining as a viable approach to landfill remediation and resource recovery within a circular economy framework.

14.10 Recommendations

Landfill mining presents both opportunities and challenges in recovering valuable materials from excavated waste. While it can reduce environmental hazards, the economic viability of the process is hindered by contamination, high treatment costs, and market demand uncertainties. To improve the effectiveness and sustainability of landfill mining, a strategic approach is needed. The following recommendations outline key areas for improvement:

- Enhancing infrastructure and technology through public-private partnerships (PPP) can attract investments in advanced biomining infrastructure while promoting research and innovation to improve waste processing efficiency, resource recovery, and leachate management.
- Addressing high upfront costs is critical, and this can be achieved by securing government subsidies, grants, and incentives to reduce financial burdens. A phased investment strategy, starting with pilot-scale projects before full-scale deployment, can ensure better risk management.
- Capacity building and skill development should be strengthened by developing specialized training programmes for waste management professionals, equipment operators, and monitoring personnel. Collaborations with educational institutions and industries can offer certification courses in biomining technologies, while workshops and knowledge-sharing platforms can facilitate the exchange of best practices.
- Strengthening policy and regulatory frameworks is essential, including implementing policies that mandate the use of mined materials across industries, reinforcing extended producer responsibility regulations, and enhancing recycling

infrastructure to promote resource recovery. Policymaker-industry collaboration can further drive sustainable waste management strategies.
- Financial viability and market demand for mined materials can be ensured by developing market linkages, establishing quality standards, and encouraging industries to adopt mined materials through policy incentives.
- Leveraging carbon credits and green finance mechanisms can enhance the profitability and long-term sustainability of biomining projects.

By implementing these measures, India can overcome operational challenges in landfill mining, enhance legacy waste management efficiency, and thereby promote a circular economy.

References

Chandana N, Goli VSNS, Mohammad A, Singh DN (2020) Characterization and utilization of Landfill-Mined-Soil-Like-Fractions (LFMSF) for sustainable development: a critical appraisal. Waste Biomass Valoriz 12(2):641–662. https://doi.org/10.1007/S12649-020-01052-Y

Cheela VRS, John M, Dubey B (2021) Quantitative determination of energy potential of refuse derived fuel from the waste recovered from Indian landfill. Sustain Environ Res 31(1):1–9. https://doi.org/10.1186/S42834-021-00097-5/TABLES/6

CPCB (2016) Solid waste generation in 46 Metrocities. Central Pollution Control Board, Ministry of Environmental, Forest and Climate Change, p 1

CPCB (2019) Guidelines for disposal of legacy waste (old municipal solid waste). https://cpcb.nic.in/uploads/LegacyWasteBiomining_guidelines_29.04.2019.pdf

Datta M, Somani M, Ramana GV, Sreekrishnan TR (2021) Feasibility of re-using soil-like material obtained from mining of old MSW dumps as an earth-fill and as compost. Process Saf Environ Prot 147:477–487. https://doi.org/10.1016/j.psep.2020.09.051

Deendayal DP, Kasinathan R (2022) Sustainable reuse potential of landfill mining waste retrieved from urban mining sites in South India. J Mater Cycles Waste Manag 0123456789:11–13. https://doi.org/10.1007/s10163-022-01506-6

Esakku S, Palanivelu K, Joseph K (2003) Assessment of heavy metals in a municipal solid waste dumpsite. In: Workshop on sustainable landfill management. https://www.researchgate.net/publication/237259524_Assessment_of_Heavy_Metals_in_a_Municipal_Solid_Waste_Dumpsite

Ghosh A, Kartha SA (2024) Landfill biomining of legacy waste dumpsites in India: process, challenges and future perspective. In: Environmental engineering and waste management, pp 377–390. https://doi.org/10.1007/978-3-031-58441-1_13

Goli VSNS, Singh DN (2023) Valorization of landfill mined plastic waste and soil-like fractions in polymer composites—a comprehensive solution for sustainable landfill mining. J Clean Prod 420:138349. https://doi.org/10.1016/J.JCLEPRO.2023.138349

Gusca J, Fainzilbergs M, Muizniece I (2015) Life cycle assessment of landfill mining project. Energy Procedia 72:322–328. https://doi.org/10.1016/j.egypro.2015.06.047

Hölzle I, Somani M, Ramana GV, Datta M (2022) Heavy metals in soil-like material from landfills—resource or contaminants? J Clean Prod 369:133136. https://doi.org/10.1016/J.JCLEPRO.2022.133136

Jain P, Powell JT, Smith JL, Townsend TG, Tolaymat T (2014) Life-cycle inventory and impact evaluation of mining municipal solid waste landfills. Environ Sci Technol 48(5):2920–2927. https://doi.org/10.1021/es404382s

Jain M, Kumar A, Kumar A (2023) Landfill mining: a review on material recovery and its utilization challenges. Process Saf Environ Prot 169:948–958. https://doi.org/10.1016/J.PSEP.2022.11.049

Kaartinen T, Sormunen K, Rintala J (2013) Case study on sampling, processing and characterization of landfilled municipal solid waste in the view of landfill mining. J Clean Prod 55:56–66. https://doi.org/10.1016/J.JCLEPRO.2013.02.036

Kaza S, Yao LC, Bhada-Tata P, Van Woerden F (2018) What a waste 2.0: a global snapshot of solid waste management to 2050. https://doi.org/10.1596/978-1-4648-1329-0

Krook J, Svensson N, Eklund M (2012) Landfill mining: a critical review of two decades of research. Waste Manag 32(3):513–520. https://doi.org/10.1016/J.WASMAN.2011.10.015

Kurian J, Esakku S, Nagendran R (2007) Mining of compost from dumpsites and bioreactor landfills. Int J Environ Technol Manag 7(3–4):317–325. https://doi.org/10.1504/IJETM.2007.015149

Mandpe A, Kumar S, Lakshmikanthan P, Hettiarachchi H (2019) Mining for recovery as an option for dumpsite rehabilitation: case study from Nagpur, India. J Environ Eng Sci 15(2):52–60. https://doi.org/10.1680/jenes.19.00021

Singh R, Budarayavalasa S (2024) Legacy waste management and dumpsite remediation in India: a policy drive for garbage-free cities. In: Solid waste management for resource-efficient systems. Elsevier. https://doi.org/10.1016/b978-0-443-23775-1.00006-0

Singh A, Chandel MK (2020) Effect of ageing on waste characteristics excavated from an Indian dumpsite and its potential valorisation. Process Saf Environ Prot 134:24–35. https://doi.org/10.1016/J.PSEP.2019.11.025

Singh A, Mankhair RV, Chandel MK (2022) Open dumps and circular economy: assessing landfill mining potential, benefits and challenges for Indian dumpsites, pp 47–63. https://doi.org/10.1007/978-3-031-07785-2_3

Somani M, Datta M, Ramana GV, Sreekrishnan TR (2018) Investigations on fine fraction of aged municipal solid waste recovered through landfill mining: Case study of three dumpsites from India. Waste Manag Res 36(8):744–755. https://doi.org/10.1177/0734242X18782393

Somani M, Datta M, Ramana GV, Hölzle I, Sundaram R, Sreekrishnan TR (2022) Effect of depth of landfill on the characteristics of soil-like material of aged waste: a case study of Bhalswa dumpsite, India. J Mater Cycles Waste Manag 24(5):1902–1912. https://doi.org/10.1007/S10163-022-01447-0/FIGURES/11

Wagner TP, Raymond T (2015) Landfill mining: case study of a successful metals recovery project. Waste Manag 45:448–457. https://doi.org/10.1016/J.WASMAN.2015.06.034

Chapter 15
Checklist-Based Framework for Implementation of Landfill Mining Projects in India

Manoj Datta, G. V. Ramana, and Mohit Somani

15.1 Introduction

Landfill mining (or Biomining, as it is often referred to in India) is a field-based practice which has evolved recently in India and differs significantly from global practices followed in developed countries. It is a rapid reclamation technique for removing legacy waste from old MSW dumps and recovering resources as well as releasing precious urban land for further use. In many major cities, where MSW dumps have reached full heights and are a visual eyesore, their "clean-up and removal" in a short time frame has been accorded the highest priority and the progress is being monitored closely by the National Green Tribunal (NGT) (Datta 2025).

After gaining experience from early field operations during the period 2018 to 2020, the efforts are now focused on segregating three major product streams from excavated waste, namely: (a) large sized, heavy C&D waste components (15–20%), (b) lightweight combustibles [plastics, textiles, and wood, etc. (10–15%)], and (c) finally gravel + soil-like-material which forms 60–80% of the mined material, other than (d) some miscellaneous rejects (Soamni et al. 2018; Datta et al. 2021). Most recyclable resources such as paper, metals, glass, etc. are already pre-recovered at the household level by an efficient informal system which has worked well in India for several decades.

Backhoe-excavators and rotating screens (trommels) are the backbone of landfill mining operations which yield high productivity and medium quality products after segregation through rotating sieves.

M. Datta (✉) · G. V. Ramana
Department of Civil Engineering, Indian Institute of Technology Delhi, New Delhi, India
e-mail: mdatta@civil.iitd.ac.in

M. Somani
School of Infrastructure, Indian Institute of Technology Bhubaneswar, Khurda, Odisha, India

M. Somani et al. (eds.), *Landfill Mining of Legacy MSW: Global Insights and Indian Experience*, Springer Transactions in Civil and Environmental Engineering, https://doi.org/10.1007/978-981-95-3959-8_15

The term "Biomining" adopted in India by the Central Pollution Control Board and the NGT implies the presence of significant quantities of biodegradables in the excavated legacy waste and their stabilization during mining operations, through inoculation with composting bio-cultures. Laboratory tests on legacy waste samples from various sites and at different depths have revealed that the organic content in legacy waste of Indian waste dumps lies below 10–15% (as against 35–45% when fresh waste was landfilled) thereby indicating that most of the biomaterials have already degraded in the early years (Somani et al. 2020, 2023; Somani 2025). Consequently, Landfill Mining in India has metamorphosed into a fast paced excavate-and-segregate operation for moist or wet-waste which minimizes the time spent on the intermediate steps of waste stabilization and drying.

15.2 Checklist-Based Framework

The planning, design, and operation of landfill mining projects are affected by a wide variety of factors ranging from site conditions, waste characteristics, product offtake opportunities, stability and safety issues, peculiarities of excavation -stabilization-segregation-storage-transportation operations, protective measures at offsite reuse locations, post-mining use of mined sites and subsurface remediation.

A one-size-fits-all set of recommendations are not suitable for various conditions encountered. For example, above-ground mounds, several tens of meters high have to be treated differently from below-ground dumps, several meters deep or even side-slope and cross-valley dump sites. Hence, a checklist-based framework which outlines the important factors/components to be addressed and the approaches to be adopted is presented in this chapter and discussed hereafter. An implementer or owner can adopt innovative approaches which are found to be suitable and efficient for the suggested framework. Over the next few years, a broad set of recommendations/guidelines may emerge, and these may be finalized and documented by the regulators.

15.3 Site Conditions and Waste Characteristics

Table 15.1 indicates the checklist for important factors relating to site conditions and waste characteristics which need to be ascertained in advance before undertaking planning and design activities. The quantum of waste, its components and the estimated quantity of potential reusable materials should be ascertained through field and laboratory studies as also the presence of leachable contaminants in the excavated material.

Table 15.1 Site conditions and waste characterization

Component	Purpose	Approach	Outcome/Benefit
Site conditions and above-ground characteristics of the site	Map the landfill's spatial and volumetric extent; assess environmental conditions	Visual inspection of site and immediate surroundings, records of waste quantities, emissions etc. drone survey, topographic survey	Assessment of environmental conditions, waste flow, approach routes etc., quantification of footprint shape and size, base area, morphology, height, estimated waste volume and weight, critical slopes and inclinations
Below-ground characteristics of the site	Estimate depth of waste and understand subsurface profile beneath waste	Borehole drilling and sampling	Depth of waste, subsoil stratigraphy, groundwater level and quality, bedrock depth
Characteristics of waste at shallow depth	Assess composition and recyclability	Excavation of pits (1–2 m wide and deep); compositional and grain size analysis of large- quantity samples	Estimate of potential recyclable materials
Variation of waste characteristics with depth	Understand vertical heterogeneity	Borehole sampling at regular intervals and lab analysis	Trends, with depth, of composition, moisture, organic content, leachable metals/salts, others

15.4 Planning and Design Considerations

The checklist for planning and design considerations is presented in Table 15.2. Planning aspects cover identification of (i) nearby industries which can reuse the lightweight combustible products of the mining project as well as (ii) local infrastructure projects which can offtake the gravel + soil-like material for earth-fills and roadworks and also (iii) construction-material processing plants which can process the C&D waste. In addition, strategies for onsite reuse or onsite storage of unutilized materials and rejects are to be finalized. Based on the desired rate of mining, a phased execution plan with multipoint locations for excavation and processing is drawn up. Provision for revision of the plan, periodically at short intervals is incorporated to take into account the variability in waste excavation conditions. Based on the estimated income from sale of reusable products and the cost of landfill mining operations, the requirements of subsidy from the government are worked out.

Design considerations focus on stability aspects of the excavation zones and processing zones as well as design of support infrastructure, internal roads, berms, surface water drainage facilities as well as fencing for site security.

Table 15.2 Planning and design considerations

Component	Purpose	Approach	Outcome/Benefit
Product demand mapping	Align excavation rates with offtake rates	Regional survey of users (cement plants, power plants, WTE plants, industries, embankment works, low-lying areas, pits, landfills)	Reuse strategy (offsite and onsite); onsite storage strategy (temporary and long-term)
Economic analysis ()	Assess financial sustainability	Arrive at unit cost of operations (per ton of waste) from nearby sites and estimated income (if any) from reuse organizations	Estimate requirements of subsidy from the government
Phased multipoint excavation plan	Manage logistics and resources to achieve desired production rates	Divide into monthly or annual excavation / mining zones (multiple locations) with sequential processing / segregation operations	Efficient mining and processing with minimal disruption
Periodic revision of excavation plan	Update size of phases on monthly / biweekly basis	Maintain daily records of excavated zones	Updated plans for progress of work
Slope stability	Ensure safe regrading of unstable areas during excavation and dumping operations	Stability analysis (local & global), regrading design	Stable working conditions, minimized waste-slide / slope-failure risk
Covers for exposed waste	Reduce odor, emissions, and erosion	Temporary soil/C&D covers over exposed or regraded slopes	Environmental protection and aesthetic improvement
Internal roads, berms	Enable efficient vehicle movement	Design of stable berms for operations and internal roads for transportation	Uninterrupted transport / removal of processed material
Surface water drainage	Efficient collection and removal of precipitation during monsoon	Garland drains with adequate gradients and capacity	Improved stability during monsoons
Infrastructure setup	Support operational activities	Installations: weighbridge, office, holding area, workshop, utilities, fencing etc.	Adequate site facilities, compliance with workplace norms

15.5 Operational Aspects

Site operations involve optimizing the size and number of equipment deployed for excavation, stabilization and drying, segregation, stockpiling/storage, loading, transportation and handling of rejects. Table 15.3 presents the checklist for operational aspects.

Table 15.3 Operational aspects

Component	Purpose	Approach	Outcome/Benefit
Excavation	Recovery of buried legacy waste	Use of excavators and loaders	Waste ready for processing
Stabilization/ drying	Accelerated stabilization and reduction in moisture of waste for improved handling	Use of windrows, applying bio-cultures, air drying, mixing, turning, etc. (this step requires adequate space and time and affects productivity)	Waste suitable for trommelling
Trommelling	Segregation of waste components	Rotating screens with appropriate mesh sizes	Separated reusable fractions—C&D material, combustibles (light weight), gravel + soil-like material, other recyclables
Storage & Transportation	Manage material flow	Designated temporary storage areas and transport logistics	Efficient material flow; optimized production rate; minimized double handling
Handling of rejects	Safe disposal of unusable material	Identification of contaminants, feasibility of onsite storage in small new lined landfill with containment.	Reduced environmental impact

Table 15.4 Safety aspects

Component	Purpose	Approach	Outcome/Benefit
Emissions monitoring	Control air and water pollution	Regular sampling for methane, dust, leachate, others	Safe working and ambient environment
Fire safety	Prevent fires in combustible waste	24 × 7monitoring; fencing of working and storage areas; fire prevention systems and adequate stocking of fire smothering materials; regular checks	Reduced fire risk, compliance with safety norms
Stability	Prevent slippages or slides of waste mass	Slope stability analysis to identify critical inclinations and heights of excavation face which can undergo failure during monsoons	Improved stability, reduced risks and timely suspension of operations during rain

15.6 Safety Aspects

During and after operations, site safety involves monitoring and controlling emissions, preventing methane induced fires and ensuring waste mass stability. Table 15.4 provides the key checkpoints.

15.7 Offsite Reuse and Protective Measures

Table 15.5 addresses the issue of reuse of mixed products and requirement of offsite protective measures. Products of landfill mining are secondary materials which are inferior to the primary materials, often mixed intricately with foreign material. The lightweight combustibles find utility in cement industry, thermal power plants, waste-to-energy plants, etc. The heavy and large sized C&D material are used as recycle aggregate, after processing in a C&D plant. The gravel and soil-like material is utilized in earth-filling of low-lying areas and embankments for roadworks, etc. Table 15.5 lists the approach for assessing the need for protective measures in vulnerable areas such as deep pits close to water table.

15.8 Post -Mining Use of Site and Remediation

Table 15.6 provides the checklist for immediate post-mining use of sites where landfill mining is completed fully or where some residual waste still remains buried and the long-term remediation measures are to be considered. For early use by public, limited access and stringent environmental monitoring is necessary. Construction of buildings and infrastructure can be considered only after stabilization and strengthening measures (in conjunction with remediation) have been undertaken, wherever needed.

Table 15.5 Offsite reuse and protective measures

Component	Purpose	Approach	Outcome/Benefit
Product reuse	Maximize resource recovery of desired quality for offsite reuse	Improving quality of segregation by better drying of waste, pulverising and extended time for tromelling. Furthermore, post-processing refining by air blowing/cleaning/shredding of lightweight combustibles.	Usable quality of combustibles/RDF for cement plants/power plants/WTE plants/industries. C&D waste for processing plants. Gravel + soil-like material for roadworks and earthworks.
Protective measures at filling sites	Ensure safe reuse at bulk earth-filling sites in vulnerable areas	Assessing possibility of spread of contaminants; end-user guidelines for adopting protective sealing-layers such as covers, liners	Protection of subsoil and ground water

Table 15.6 Post-mining use of site and subsurface remediation

Component	Purpose	Approach	Outcome/Benefit
Site redevelopment (immediate)	Convert mined area for public or private use	Feasibility study; emission monitoring and control; compacted soil cover and drainage	Green zones, parks, solar farms, parking areas, golf courses, etc.; or construction area (only after stabilization/ remediation)
Site restrictions (immediate)	Limit exposure to hazards	Limited access at start (only short daylight hours and limited locations/ zones); extend gradually after extensive geoenvironmental monitoring for few years	Short-term and long-term safety, regulatory compliance
Construction of buildings on site	Establish feasibility of developing infrastructure a few years after end-of-mining; monitoring of emissions and soil movements should indicate that stable conditions have been established	Ground strengthening / stabilization/remediation measures, protection of foundations from deleterious salts; environmental control provisions	Site suitability for built environment on stable and remediated ground
Delineation of impacted zone	Understand extent of subsurface contamination	Geotechnical and geoenvironmental field investigations; modeling	Basis for design
Soil and groundwater remediation	Remove or mitigate contamination	Remediation planning: comparison of technology options; selection; field trials; execution; monitoring	Improved environmental quality, risk reduction

15.9 Concluding Remarks

Based on the experience gained over the last 7 years in India, a checklist-based framework is suggested for implementing landfill mining projects. A wide variety of factors have been earmarked for consideration in the planning, design, and implementation stages. This framework will be useful to all those involved in execution of landfill mining projects in the country.

References

Datta M (2025) Insights from landfill mining of old MSW dumps in India; webinar on 27th March 2025 at the international webinar series on Geoenvironmental engineering, sustainability, and resiliency hosted by University of Illinois Chicago, USA

Datta M, Somani M, Ramana GV, Sreekrishnan TR (2021) Feasibility of re-using soil-like material obtained from mining of old MSW dumps as an earth-fill and as compost. Process Saf Environ Prot 147:477–487. https://doi.org/10.1016/j.psep.2020.09.051

Somani M (2025) Overcoming dumpsites rehabilitation challenges in India: landfill mining versus microbial methane oxidation systems. Indian Geotech J:1–45. https://doi.org/10.1007/s40098-025-01356-7

Somani M, Datta M, Ramana GV, Sreekrishna TR (2020) Contaminants in soil-like material recovered by landfill mining from five old dumps in India. Process Saf Environ Prot 137(2020):82–92. https://doi.org/10.1016/j.psep.2020.02.010

Somani M, Harbottle M, Datta M, Ramana GV, Sreekrishnan TR (2023) Identification and assessment of appropriate remediation management techniques for the recovery of soil-like material produced in landfill mining. J Environ Manag 348:119300. https://doi.org/10.1016/j.jenvman.2023.119300

The manufacturer's authorised representative in the EU is Springer Nature Customer Service Centre GmbH, Europaplatz 3, 69115 Heidelberg, Germany. If you have any concerns regarding our products, please contact ProductSafety@springernature.com

Printed and bound by CPI Group (UK) Ltd, Croydon, CR0 4YY
07/10/2026
02229357-0004